SAMPLE CONTROLLED THERMAL ANALYSIS

Hot Topics in Thermal Analysis and Calorimetry

Volume 3

Series Editor:
Judit Simon, *Budapest University of Technology and Economics, Hungary*

The titles published in this series are listed at the end of this volume.

Sample Controlled Thermal Analysis

Origin, Goals, Multiple Forms, Applications and Future

Edited by

O. Toft Sørensen

Risø National Laboratory, Roskilde, Denmark

and

J. Rouquerol

MADIREL, CNRS-Université de Provence, Marseille, France

KLUWER ACADEMIC PUBLISHERS

DORDRECHT / BOSTON / LONDON

A C.I.P. Catalogue record for this book is available from the Library of Congress.

ISBN 1-4020-1563-1

Published by Kluwer Academic Publishers,
P.O. Box 17, 3300 AA Dordrecht, The Netherlands.

Sold and distributed in North, Central and South America
by Kluwer Academic Publishers,
101 Philip Drive, Norwell, MA 02061, U.S.A.

In all other countries, sold and distributed
by Kluwer Academic Publishers,
P.O. Box 322, 3300 AH Dordrecht, The Netherlands.

Printed on acid-free paper

Printed in the Netherlands.

Contents

Chapter 7:
SCTA AND CATALYSIS
E.A. Fesenko, P.A. Barnes and G.M.B. Parkes

Chapter 8:
SCTA IN THE FUTURE
M. Reading

List of Contributing Authors and Laboratories

E.A. Fesenko, G.M.B. Parkes and P.A. Barnes
Centre for Applied Catalysis, University of Huddersfield, UK

E.L. Charsley
Centre for Thermal Studies, University of Huddersfield, UK

J.M. Criado and L.A. Pérez-Maqueda
Instituto de Ciencia de Materiales de Sevilla, Centro Coordinado C.S.I.C.- Universidad de Sevilla; Sevilla, Spain

P. Llewellyn, F. Rouquerol and J. Rouquerol
MADIREL Laboratory, CNRS-Université de Provence, Marseille, France

O. Toft Sørensen
Materials Department, Risø National Laboratory, Roskilde, Denmark

M. Reading
PTME, Loughborough University, Loughborough, UK

Preface

The reason for this book was the editors' feeling, actually shared by the other contributors, that a gap should be filled now. This is indeed the first book devoted to the developing branch of thermal analysis called, for 6 years, "Sample Controlled Thermal Analysis" (SCTA), where a feed-back from the sample is used to control its heating or cooling. In this method, it is therefore the sample itself which determines its own heating and cooling conditions as the name implies. This approach can be considered as a real breakthrough in the field of thermal analysis. It has indeed the advantage, as compared with conventional thermal analysis, of providing: elimination of uncertainties due to thermal effects in the sample container, improved resolution, accurate determination of reaction temperatures and accurate kinetic data. SCTA has since its introduction in the early seventies been used in many studies both on inorganic and to a certain extent organic (polymers) compounds with the aim of studying the temperature, the type and the kinetics of reactions taking place during heating and cooling: in the case of ceramics and adsorbents SCTA has even been used in synthesis of materials with specific properties. These techniques are now also available in commercial thermal analysis instruments.

The editors wish to thank the contributors to this book for their enthusiastic and ardent work. They also wish to acknowledge all colleagues throughout the world which, by their many inspiring publications and private communications, have made this book possible.

Finally it is the editors hope that this book will be an inspiration in future applications and in the development of Sample Controlled Thermal Analysis.

Ole Toft Sorensen and Jean Rouquerol

Chapter 1

GENERAL INTRODUCTION TO SAMPLE-CONTROLLED THERMAL ANALYSIS (SCTA)

J. ROUQUEROL[1] and O. TOFT SORENSEN[2]

1-Madirel Laboratory, CNRS-Université de Provence, Marseille, France
2-Risoe National Laboratory, Roskilde, Denmark

1.1. Spirit and Definition of SCTA

For a majority of technical applications, it is of upmost importance to know the behaviour of materials as they are heated or cooled. This is the objective of thermal analysis. In the beginning (say, one century ago) the sample was simply submitted to the heating from a gas burner and, later, from an electrical furnace, operated at *ca* constant power (as it was the case with the "micro-DTA" equipment marketed in the early sixties). This meant a certain lack of control on the heating rate and on the total duration of the experiment. It was therefore considered as a great improvement to impose a temperature programme to the sample, most often with a constant heating rate, *i.e.* with a constant dT/dt (where T is the sample or furnace temperature and where t is the time): the great majority of thermal analysis equipments are, nowadays, operated in this way.

Now, specially with the progress of microprocessor-based heating controllers, it has become easier and easier to decide, at will, on more sophisticated modes of operation of a thermal analysis experiment, and to adapt them to the specific material or transformation under study: this has lead to the new branch of Thermal Analysis now known as "Sample Controlled Thermal Analysis", which is the subject of the present book.

More precisely, the term "Sample-Controlled Thermal Analysis" designates the branch of thermal analysis where a feed-back from the sample is used to control its heating or cooling.

Two important features of SCTA are therefore: (i) the use of a feed-back from the sample and (ii) the absence of any pre-determined temperature programme.

This is basically different from what has been, for a long time, the only branch of thermal analysis and which we may therefore call "Conventional Thermal Analysis" (or also "Temperature-Controlled Thermal Analysis"), where the sample (or furnace) temperature is brought to follow a pre-determined programme which does not make use of any response from the sample. Nevertheless, both branches above are part of Thermal Analysis, which embraces all techniques where a property of a sample is recorded vs. temperature.

In the above definitions, the word "sample" designates the object under study; it is usually (although not necessarily) in a dense state (solid or liquid) and in the presence of a gas (or in vacuum). Among several others, the term "Sample Controlled Thermal Analysis", with the acronym "SCTA", is the one which was finally selected during the Second International Workshop organized in this field (The first one was in 1994, in the scope of ESTAC VI in Grado, Italy, and the second one in 1996, in the scope of ICTAC-11 in Philadelphia, USA). It was indeed generally felt among the participants that this term was the closest to the definition given above, which carries the "spirit" of this relatively new branch of thermal analysis. A further advantage of this term is that it had never been used earlier by any of the pioneers of this approach, so that there is no possible confusion with any particular form of it.

Concerning the "spirit" of SCTA, it can also be said that, here, the sample is not heated in a "blind" way, but after taking its response into account. This heating mode can even be said to be "respectful" of the sample, specially when the rate of transformation is kept at a value low enough to avoid, for instance, any unwanted cracks due to internal gas overpressure. The same low controlled rate is also often found desirable simply to reduce at will the unavoidable temperature gradients which are due, in great part, to the self-cooling (or self-heating) of the sample during its transformation. In case a gas phase is produced or consumed by the transformation, the pressure gradients within the sample can be reduced in the same way.

As it will be commented later in this chapter and illustrated in the whole book, this close control provided by SCTA usually results in an *enhanced resolution of the thermal analysis curve*, in a *highly homogeneous thermal treatment or preparation* and in the possibility of *meaningful kinetic studies*.

1.2. History of SCTA

Because SCTA is a still developing branch of thermal analysis and because its history is a recent one — most inventors and developers of SCTA are still alive and active at the time of the writing of this book — we shall be careful to avoid any controversy. For that purpose, we shall quote the main publications, together with the name given by each author to his technique or procedure and shall later propose (in Chapter 2) a general framework able to accommodate these many developments.

The first form of SCTA was probably the one proposed by C.S. Smith, in the US, in 1940 [1], although he was not at all conscious of opening a new general branch of thermal analysis. His "Simple Method of Thermal Analysis Permitting Quantitative Measurements of Specific and Latent Heats" was devised for metallurgic purposes. As Smith writes, it "consists in placing the specimen with its thermocouple in a refractory container of low thermal conductivity and placing this in a furnace the temperature of which is maintained a constant amount above or below the specimen temperature. This is easily managed by opposing the electromotive force of the specimen couple to that of a couple at the outside of the container, and feeding the resultant electromotive force to an automatic temperature controller of ordinary design". Here, the feed-back from the sample is the heat-flow received or delivered by the sample and which is maintained at a *ca.* constant rate, whereas the sample temperature does not follow any pre-determined programme: we are well in the scope of SCTA, as defined *ca.* sixty years later! Now, this work was only noticed by a small number of metallurgists, who used what they called the "Smith Thermal Analysis Technique", and it remained unknown by the broad community of thermal analysis until one of us was kindly introduced to it, in 1989, by F.H. Hayes, an English metallurgist [2]. Let us point out that Smith effectively built an equipment and published results obtained with a 86g sample of brass and a 150g sample of copper-silicon alloy. The sensitivity was low (the constant temperature difference between the furnace and sample was of 60K and the constant heat flow as high as 4 watts!) but the principle was there. In the proposal which we make, in the next section, for a consistent nomenclature, Smith's technique can be considered as the first form of Constant Rate DTA (CR-DTA): the equipment makes indeed use of one differential thermocouple (like in standard DTA), one of which is immersed in the sample itself.

The next forms of SCTA were independently proposed, in Budapest and in Paris, in the early sixties.

It was indeed in 1962 that L. Erdey, F. Paulik and J. Paulik filed a Hungarian patent on "A device for the automatic control of the heating programme of thermobalances in the case of a stepwise isothermal heating" (translated from Hungarian). The aim of this invention was "to vary the temperature of the electrical furnace in such a way as to allow the thermal reaction to take place, in practice, with the low rate of decomposition desired, by approaching the desired value either from the upper side or from the lower one ...” (translated from Hungarian). No special name was then given to this procedure, which we would now call Controlled Rate Thermo Gravimetry (CR-TG). This patent was published in 1965 [3].

Meanwhile and in the scope of his doctorate (which he prepared between 1959 and 1964, in the Sorbonne, in Paris) J. Rouquerol tried to improve the thermal preparation of porous alumina and beryllia (when obtained by thermal treatment of a hydroxide precursor) by controlling their heating from a signal related to the rate of water vapour evolution. He was lucky to set up the appropriate equipment and to have it working. All samples studied in his thesis were prepared by this method [4]. In 1964 he published a paper [5] entitled “Method of thermal analysis under low pressure and at constant rate of decomposition” (translated from French) where he described his set-up, based on what we would now call the Evolved Gas Detection technique. He reports application to the dehydration of a number of hydrates and hydroxides, pointing out that the same method and experiment can be used for thermoanalytical purposes and for the sake of thermal preparation under controlled conditions such as, for instance, promote the development of the specific surface area and delay the sintering. It seems that, until 1971, this technique, which we would now call Controlled Rate Evolved Gas Detection (CR-EGD) was the only which, after the publication of Smith’s paper in 1940, lead to the publication of experimental results in the field of SCTA [6–10].

It is only in 1970 that J. Rouquerol understood that the technique he was using was part of a new, much more general, method of thermal analysis which he proposed to call “Constant Rate Thermal Analysis”, with the acronym of CRTA [11]. He pointed out that “The proposed principle consists in carrying out a thermal analysis where the controlled parameter is not the temperature but a quantity directly related to the rate of decomposition and which can be, for instance:

- the flow of evolved gas (either pure or mixed with a carrier gas)
- the rate of mass variation, as obtained by derivative thermogravimetry
- the heat flux received, as given by a DTA device or by a more elaborate calorimetric device”(translated from French).

The “CRTA method” as proposed and defined in this paper, was, therefore, from the start, quite general and was never intended to be connected with any specific thermoanalytical technique.

The foregoing illustrates the fact that the concept of SCTA, which can be considered as the introduction, in thermal analysis, of an elaborate automation relying on a feed-back from the sample, was "in the air" of our technological time and germinated, quite independently, in several minds quite far from each other either geographically or scientifically.

Further novelties and landmarks in the development of SCTA are briefly listed hereafter, before being developed in the rest of the book:

1971: J. and F. Paulik publish the first experimental results obtained with the method patented in 1962 and which they call "quasi-isothermal thermogravimetry" or also "quasi-isothermal, quasi-isobaric thermogravimetry" when improving it by the use of their labyrinth crucible [12]. As they later wrote, they "made the first attempt in 1962 to develop this technique, but it was only in 1971 that every obstacle to the realization of an appropriate measuring system could be eliminated" [13].

1972: F. and J. Rouquerol propose the first kinetical application of SCTA with the "Rate-Jump Method" to determine the activation energies all along the thermal decomposition of a single sample, by periodically switching the imposed rate of decomposition between two pre-set values [14].

1973: F. and J. Paulik develop the analytical applications of SCTA by taking advantage of the various simultaneous measurements provided by their "Derivatograph" carrying out simultaneous TG, DTG and DTA [15].

1977: J. and F. Paulik, by modifying their thermobalance into a thermodilatometer and by keeping the quasi-isobaric conditions, are able to carry out a "Thermal dilation study under quasi-isothermal, quasi-isobaric conditions" with good selectivity and resolution [16]; this enters the present scope of Controlled Rate Thermo Dilatometry (CR-TD).

1978–1980: O.T. Sorensen proposes a new form of SCTA, specially suited for kinetical applications, which he later called "Stepwise Isothermal Analysis" (SIA) [17] and which aims to take advantage from both SCTA and Isothermal Kinetics. In this approach, indeed, SCTA proper is used to bring the sample at the appropriate temperature, able to provide the desired order of magnitude of the reaction rate, whereas the procedures and concepts of isothermal kinetics are fully applicable during the isothermal periods which are introduced between two successive steps. O.T. Sorensen applies this approach both to the case of thermogravimetry [18] and to that of thermodilatometry [19].

1985: M.H. Stacey extends the application of SCTA to the study of reactions of solids with gases like the reduction of copper or nickel catalysts in a flow of H_2 and He, using a fluidized bed under atmospheric pressure [20].

1992: M. Reading explicitely proposes to take into account complex heating programmes and proposes to speak of “Constrained Thermal Analysis” [21].

1992: Sauerbrunn et al. introduce the use, in SCTA, of a complex algorithm for the heating control, which involves at the same time the rate of transformation, the temperature and the time and which aims at reducing the duration of the experiment, though preserving the resolution provided by SCTA [22].

1994: J.M. Criado and A. Ortega propose a form of SCTA where the rate of reaction is steadily accelerated, which increases the discriminating power of SCTA between several mechanisms of reaction [23].

1994–1995: P. Barnes, G.M.B. Parkes, D.R. Brown and E.L. Charsley develop a versatile Evolved Gas Analysis equipment and software able to be run under various modes of SCTA, specially SIA and CRTA with various heating algorithms [24,25].

1996: J. Rouquerol and J.M. Fulconis modify the Rate-Jump method into a method with a sinusoidal variation of the transformation rate (Modulated SCTA) which results into a corresponding sinusoidal variation of the sample temperature and allows the periodical, asumptionless, determination of the activation energy [26]. This is applied to experiments of Sample-Controlled EGD.

1997: R.L. Blaine develops the TG version of modulated SCTA [27].

1998–2000: G.M.B. Parkes, P.A. Barnes, E.L. Charsley, M. Reading and I. Abrahams examine the possibilities and merits of SCTA experiments based on heating algorithms which make use of the “peak shape” (in case, for instance, of an EGA or DTG type signal). They distinguish between “responsive” peak-shape SCTA methods and “predictive” peak-shape SCTA methods [28,29].

References

1. C.S. Smith, Trans. A.I.M.E. (Metal Division) 137 (1940) 236.
2. F.H. Hayes, UMIST, Manchester, UK, personal communication.
3. L. Erdey, F. Paulik and J. Paulik, Hungarian Patent No. 152197, registered 31 October 1962, published 1 December 1965.
4. J. Rouquerol, Thesis, Faculty of Sciences of Paris University, 19 November 1964 (Série A, No. 4348, No. d’ordre 5199).
5. J. Rouquerol, Bull. Soc. Chim. Fr. (1964) 31.
6. J. Mayet, J. Rouquerol, J. Fraissard and B. Imelik, Bull. Soc. Chim. Fr. (1966) 2805.

7. A. Baumer and M. Ganteaume, C.R. Acad. Sci. 266 (1968) 120.
8. J. Rouquerol in Thermal Analysis, in Thermal Analysis, Academic Press, New York 1 (1969) 281.
9. G. Chottard, J. Fraissard and B. Imelik, Bull. Soc. Chim. Fr. (1967) 4331.
10. J. Kermarec, J. Fraissard, J. Elston and B. Imelik, J. Chim. Phys. 65 (1968) 920.
11. J. Rouquerol, J. Therm. Anal. 2 (1970) 123.
12. J. Paulik and F. Paulik, Anal. Chim. Acta 56 (1971) 328.
13. J. Paulik and F. Paulik in G. Svehla (Eds.) Comprehensive Analytical Chemistry, Elsevier, Amsterdam, Vol. XII, Part A (1981) 48.
14. F. Rouquerol and J. Rouquerol in H.G. Widemann (Eds.) Thermal Analysis, Vol. 1, Birkhäuser, Basel (1972) 373.
15. F. Paulik and J. Paulik, Anal. Chim. Acta 67 (1973) 437.
16. J. Paulik and F. Paulik Thermal Analysis, Proc. 5th ICTA Conference, Kyoto, Heyden and Sons, London (1977).
17. O.T. Sorensen, J. Therm. Anal. 13 (1978) 429.
18. O.T. Sorensen, Thermochim. Acta 50 (1980) 163.
19. O.T. Sorensen in Proc. 5th Meeting of AICAT, AICAT, Trieste (1983) 25.
20. M.H. Stacey, Anal. Proc. 22 (1985) 242.
21. M. Reading, in "Thermal Analysis, Techniques and Applications" E.L. Charsley and S.B. Warrington (Eds.), Royal Society of Chemistry, Cambridge (1992) 126.
22. P.S. Gill, S.R. Sauerbrunn and B.S. Crowe, J. Thermal Analysis, 38 (1992) 255.
23. A. Ortega, L.A. Pérez-Maquéda and J.M. Criado, J. Thermal Anal. 42 (1994) 551.
24. P.A. Barnes, G.M.B. Parkes and E.L. Charsley, Anal. Chemistry, 66 (1994) 2226.
25. P.A. Barnes, G.M.B. Parkes, D.R. Brown and E.L. Charsley, Thermochimica Acta, 269/270 (1995) 665.
26. J. Rouquerol and J.M. Fulconis, ICTAC 11 Book of Abstracts, P. Gallagher (Ed.), Philadelphia (1996) 272.
27. R.L. Blaine, Proceedings of NATAS Meeting, R.J. Morgan (Ed.) (1997) 485
28. M. Reading in: M.E. Brown (Ed.), Handbook of Thermal Analysis and Calorimetry, Elsevier, 1998 (Chapter 8).
29. G.M.B. Parkes, P.A. Barnes, E.L. Charsley, M. Reading and I. Abrahams, Thermochimica Acta, 354 (2000) 39.

Chapter 2

A FRAMEWORK FOR THE SCTA FAMILY

J. ROUQUEROL[1] and O. TOFT SORENSEN[2]

1-Madirel Laboratory, CNRS-Université de Provence, Marseille, France
2-Risoe National Laboratory, Roskilde, Denmark

2.1. Representing the Specificity of SCTA

Figure 2-1 provides a simple representation of thermal analysis in general, where a physical property "X" of sample "S" is recorded vs. temperature "T", as the sample is heated or cooled. As we know, this physical property can be a mass, a heat content, a length, or any other mechanical, electrical or optical property. Because this is a general representation, encompassing any type of thermal analysis, nothing is indicated about the possible modes of heating control.

Now, two modes of heating are basically different from each other and clearly introduce the specificity of SCTA. They are represented in Figures 2-2 and 2-3.

Figure 2-2 represents the common mode of heating control, where the control loop makes use of a *temperature* signal (temperature of the sample proper or of its surroundings). This signal is combined with time in the heating control algorithm (most often able, nowadays, to provide proportional, derivative and integral effects) which itself acts on the power unit and, therefore, on the heating proper, in order to impose *a* predetermined temperature programme. This is what can be called "Temperature-Controlled Thermal Analysis" or also, since it is the most widely used, "Conventional Thermal Analysis".

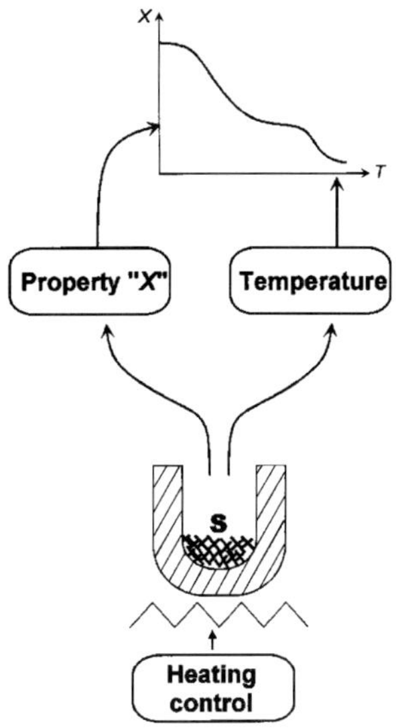

Figure 2-1. Thermal analysis in general.

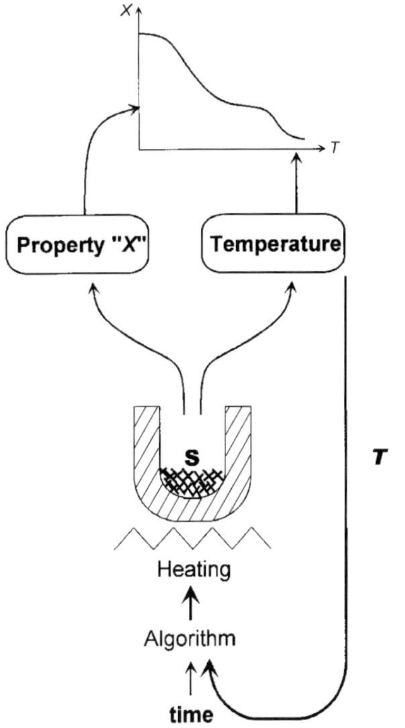

Figure 2-2. Temperature-controlled thermal analysis.

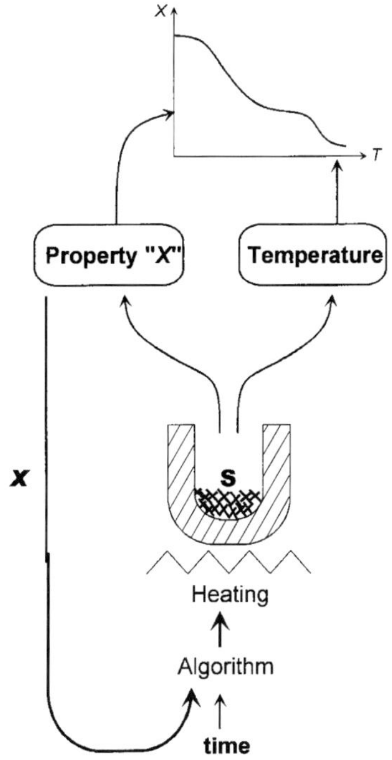

Figure 2-3. Rate-controlled thermal analysis.

Now, Figure 2-3 shows a fully different situation. Here, the heating control algorithm receives a feedback from the sample, which is a signal related to any physical property "X" influenced by the thermal transformation of the sample. Since this signal is related to the advancement of the transformation and also, once the time is also measured, to the *rate of the transformation,* this can be called "Rate-Controlled Thermal Analysis". What is programmed, here, is not any more the temperature (whose profile is unknown in advance) but the rate of the thermal transformation. This approach contains the typical feature of SCTA (which is, as stated in the first page of Chapter 1, the use of a *feed-back* from the sample to control its heating or cooling) and can be considered as the simplest form of SCTA.

The combination of the two above heating control loops is also possible and leads to a more general form of SCTA, represented in Figure 2-4, which can now be called "Rate- and Temperature-Controlled Thermal Analysis".

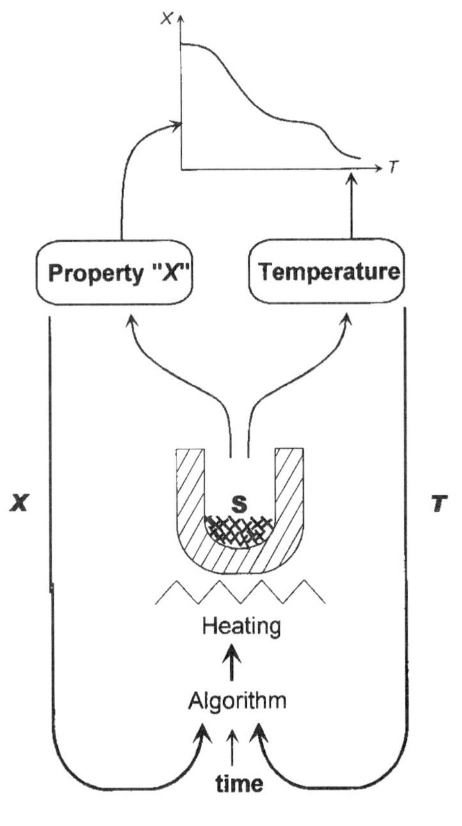

Figure 2-4. Rate- and temperature-controlled thermal analysis ("Generalized SCTA").

The heating control algorithm of this generalized form of SCTA (Reading's "Constrained Rate Thermal Analysis" [1]) can be written as:

$$P = f(X, T, t) \tag{1}$$

where P is the heating power, X is one (or several) properties of the sample related to its thermal transformation, T is the sample temperature (or a surrounding temperature) and t is the time.

2.2. Simple Distinction Between the Various Forms of SCTA

Since a major interest of the SCTA approach is the possibility of directly controlling the rate of the thermal reaction, SCTA is naturally divided into two major branches: one where the heating of the sample is exclusively rate-controlled (only the left-hand control loop, like in Fig. 2-3) and one where the heating is controlled both by the rate (or by the simple occurrence) of the thermal transformation and by the sample temperature, as represented in Figure 2-4. The "simple occurrence" means that, although a feed-back from the sample is still used, it can be of short duration and act upon the heating in a discontinuous way (for instance: signal corresponding to a sudden structural change of the sample). Each of these two major branches of SCTA can itself be easily split into two subdivisions, as shown in Table 2-1.

Table 2-1. The main forms of SCTA

A – Solely ***Rate***-Controlled SCTA	B – ***Rate- and Temperature***-Controlled SCTA
A-1 – ***Constant*** Rate TA	B-1 – ***Alternate*** Rate- and Temperature-Controlled
A-2 – ***Varying*** Rate TA	B-2 – ***Combined*** Rate and Temperature-Controlled

This provides us with four, clearly separated, types of SCTA, which will be used, hereafter, as headings for the listing of the various approaches to SCTA which were tried until now. Each one is quoted with the name proposed by its author (*in italics*) followed by the bibliographical reference where this name was first proposed. Then, for general consistency, we finally give in brackets the name which best fits, nowadays, the general framework able to incorporate all SCTA techniques. A detailed presentation of most of these various forms of SCTA can be found in Chapter 3.

A-1 Constant Transformation Rate Thermal Analysis (Solely Rate-Controlled)

All techniques listed hereafter have in common that they try to *keep the rate of transformation around one preset value*. This is achieved more or less accurately, depending on the technical possibilities and on the needs, but the basic, common, principle is there. Most of these experiments try to take advantage of the high resolution provided by the selection of low transformation rates; this explains, for a number of them, the choice of a name beginning by "Quasi-Isothermal...". Now, the expression "Constant Transformation Rate Thermal Analysis" is more general and also embraces the cases where the selection of a high rate is considered to be interesting, for instance in order to study the diffusion limited regime of a reaction (4). Finally, let us stress that we include in this section experiments which may start and/or end with a conventional temperature programme, whereas the central, interesting part of the experiment is aimed to be carried out at a constant transformation rate. An initial constant heating rate step may indeed be needed to reach the temperature where the thermal transformation starts, whereas a final isothermal step may be requested for some preparative purposes.

- "*Simple method of Thermal Analysis Permitting Quantitative Measurements of Specific and Latent Heats*", Smith, 1940 [2] (Constant Rate DTA).
- "*Constant Rate TA*", Rouquerol, 1964 [3] (Constant Rate EGD under vacuum).
- "*Device for the automatic control of the heating programme of thermobalances*", Erdey, Paulik and Paulik, 1965 [8] (Constant Rate TG).
- "*Rate controlled sintering*", Palmour and Johnson, 1967 [4] (Constant Rate Thermodilatometry).
- "*Vacuum thermal analysis apparatus, with controlled residual pressure and with constant decomposition rate ... with simultaneous measurement by thermogravimetric analysis*", Rouquerol, 1969 [5] (Simultaneous TG and Constant Rate EGD under vacuum).
- "*Constant Rate Thermal Analysis (CRTA)*", Rouquerol, 1970 [6] (Constant Rate Thermal Analysis).
- "*Quasi-Isothermal, Quasi-Isobaric TG*", Paulik and Paulik, 1971 [7] (Constant Rate TG under constant pressure of self-generated atmosphere).

- "*Simultaneous measurement of a thermal flow and of a gas flow by means of Constant decomposition Rate Thermal Analysis*", Ganteaume and Rouquerol, 1971 [9] (Simultaneous DSC and Constant Rate EGD under vacuum).
- "*Quasi-Isothermal, Quasi-Isobaric Thermodilation*", Paulik and Paulik, 1977 [10] and "*Quasi-Isothermal Dilatometry (QID)*", Sorensen, 1978 [11] (Constant Rate Thermodilatometry).
- "*Quasi-Isothermal DTA (Q-DTA)*", Paulik, Paulik and Arnold, 1985) [12] (Constant Rate DTA).
- "*Constant Rate Thermal Analysis (CRTA) in a fluidised bed reactor*", Stacey, 1981 [15], "*Controlled Rate Thermal Analysis using an Infra-Red Gas Analyser*", Reading, 1985 [18], "*Constant Rate Thermal Analysis (CRTA) under flow of inert or reactive gases*", Arellano, Criado and Real, 1988 [14], Real, Alcala and Criado, 1992 [17], "*High Resolution EGA using Rate-Controlled Thermal Analysis*", Barnes, Parkes, Brown and Charsley [19] 1995, "*CRTA with a carrier gas*" Rouquerol and Fulconis [20] 1996 (Constant Rate Evolved (or Reacted) Gas Analysis under flow of inert or reactive gas).

A-2 Varying Transformation Rate Thermal Analysis (Solely Rate-Controlled)

- "*Rate-Jump Thermal Analysis*", Rouquerol and Rouquerol, 1972 [21] (Modulated Rate Thermal Analysis).
- "*Controlled Decomposition Rate Evolved Gas Analysis*", Thevand, Rouquerol and Rouquerol, 1982 [22] (Controlled Rate EGA).
- "*Increasing Rate Thermal Analysis*" Ortega, Perez-Maqueda and Criado, 1994 [23] (Increasing Rate Thermal Analysis).
- "*Modulated Rate EGD*", Rouquerol, 1996 [20] (Modulated Rate EGD) "*Modulated Rate TG*", Blaine, 1997 [24, 25] (Modulated Rate EGD).
- "*Responsive or predictive "peak-shape" SCTA*" (Parkes, Barnes, Charsley, Reading and Abrahams, 2000 [26] (Controlled Rate Thermal Analysis with permanent adaptation of rate to peak shape).

B-1 Alternate Transformation Rate- and Temperature-Controlled Thermal Analysis

- "*Stepwise Isothermal Analysis (SIA)*", Sorensen, 1983 [27] (Alternate Isothermal and Rate Controlled Thermal Analysis).
- "*Forced Stepwise Isothermal Analysis (FSIA)*", Sorensen, 1992 [28] (Alternate Isothermal and Rate Controlled Thermal Analysis, with splitting of the isotherms).

- "*Max Res*" (Schenker and Riesen, 1997 [29] (Controlled *transformation* Rate TG with two selected *heating* rates).

B-2 Combined Transformation Rate- and Temperature-Controlled Thermal Analysis

Here, the transformation rate is still a controlling parameter but it is not programmed in advance, since it is connected, in a more or less complex or loose way, to temperature and time; this is what M. Reading called "Constrained Rate TA" [1].

- "*High Resolution Thermogravimetry*" or "*Hi-Res TGA*" (Sauerbrunn, Gill and Crowe, 1992 [30] (Combined Rate-Controlled and Temperature-Controlled TG).
- "*Dynamic Heating Rate*" (Opfermann [31] (Combined Rate-Controlled and Temperature-Controlled TG).

2.3. Concluding Remarks

One may be amazed by the variety of set-ups and techniques already proposed in a form which directly enters the scope of SCTA. In spite of being relatively recent, this approach has indeed been extremely inspiring and we shall see, at the end of this book, that novelties and inventions in this field still have a future. The framework just proposed completely relies on the specific character of SCTA, *i.e.* the special heating control which makes use of a feed-back from the sample. As we saw, it allows to distinguish four main types of SCTA with practically no overlapping, so that we may hope that, for a while, it will also be able to accommodate most of the forthcoming forms of SCTA.

References

1. M. Reading, in "Thermal Analysis, Techniques and Applications" E.L. Charsley and S.B. Warrington (Eds.), Royal Society of Chemistry, Cambridge (1992) 126.
2. C.S. Smith, Trans. A.I.M.E. (Metal Division) 137 (1940) 23.
3. J. Rouquerol, Bull. Soc. Chim. Fr. (1964) 31.
4. H. Palmour and D.R. Johnson, in "Sintering and Related Phenomena", G.C. Kuczynski, N.A. Hooton and C.F. Gibbon (Eds.), Gordon and Breach, New York (1967) 779.
5. J. Rouquerol, J. Therm. Anal. 1 (1969) 281.
6. J. Rouquerol, J. Therm. Anal. 2 (1970) 123.

7. J. Paulik and F. Paulik, Anal. Chim. Acta 56 (1971) 328.
8. L. Erdey, F. Paulik and J. Paulik, Hungarian Patent No. 152197, registered 31 October 1962, published 1 December 1965.
9. M. Ganteaume and J. Rouquerol, J. Therm. Anal. 3 (1971) 413.
10. J. Paulik and F. Paulik, Proc. 5th Int. Conf. Thermal Analysis (Kyoto), Heyden, London, 1 (1977) 75.
11. O.T. Sorensen, J. Therm. Anal. 13 (1978) 429.
12. J. Paulik, F. Paulik and M. Arnold, Hungarian Patent No. 194.405/1985.
13. J. Paulik, "Special Trends in Thermal Analysis", John Wiley and sons, Chichester, New York (1995) p. 232.
14. J. Arellano, J.M. Criado and C. Real, Thermochimica Acta 134 (1988) 365.
15. M.H. Stacey in "Proceedings 2nd European Symposium on Thermal Analysis" (Aberdeen), D. Dollimore (Ed.), Heyden, London, 408.
16. M.H. Stacey, Anal. Proc. 22 (1985) 242.
17. C. Real, M.D. Alcala and J.M. Criado, J. Therm. Anal. 38 (1992) 797.
18. M. Reading and J. Rouquerol, Thermochimica Acta 85 (1985) 299.
19. P.A. Barnes, G.M.B. Parkes, D.R. Brown and E.L. Charsley, Thermochimica Acta, 269/270 (1995) 665.
20. J. Rouquerol and J.M. Fulconis, ICTAC 11 Book of Abstracts, P. Gallagher (Ed.), Philadelphia (1996) 272.
21. F. Rouquerol and J. Rouquerol in "Thermal Analysis", H.G. Widemann, (Ed.), Vol. 1, Birkhäuser, Basel, Vol. 1, (1972) 373.
22. G. Thevand, F. Rouquerol, and J. Rouquerol in "Thermal Analysis", B. Miller, (Ed.), John Wiley and Sons, NewYork, Vol. 2, (1982) 1524.
23. A. Ortega, L.A. Pérez-Maquéda and J.M. Criado, J. Therm. Anal. 42 (1994) 551.
24. R.L. Blaine, Proceedings of NATAS Meeting, R.J. Morgan (Ed.) (1997) 485.
25. R.L. Blaine, American Laboratory, January 1998, 21.
26. G.M.B. Parkes, P.A. Barnes, E.L. Charsley, M. Reading and I. Abrahams, Thermochimica Acta, 354 (2000) 39.
27. O.T. Sorensen in Proc. 5th Meeting of AICAT, AICAT, Trieste (1983) 25.
28. O. Toft Sorensen, J. Thermal Analysis, 38 (1992) 213.
29. B. Schenker and R. Riesen, USER COM, Information for users of Mettler Toledo Thermal Analysis Systems, December 1997, 10.
30. P.S. Gill, S.R. Sauerbrunn and B.S. Crowe, J. Therm. Anal., 38 (1992) 255.
31. J. Opfermann, J. Netzsch Geraetebau, Germany-private communication.

Chapter 3

BASIC SCTA TECHNIQUES

J. ROUQUEROL[1] and O. TOFT SORENSEN[2]

with contributions from

P. BARNES[3], E.L. CHARSLEY[3], E. FESENKO[3] and M. READING[4]

1-Madirel Laboratory, CNRS-Université de Provence, Marseille, France
2-Risoe National Laboratory, Roskilde, Denmark
3-Materials Department, University of Huddersfield, UK
4-IPTME, Loughborough University, Loughborough, UK

3.1. Introduction

Our aim, in this chapter, is to introduce the reader to the principle of operation, measurement and, sometimes also, data processing concerning the basic SCTA techniques used until now. For the sake of clarity, we shall follow the same order as in our listing of SCTA techniques in Chapter 2, i.e. based, as much as possible, both on logic and chronology. The headings of the paragraphs will use the simplest names suggested by the framework of Chapter 2, but the titles initially used by each author will also be reminded. We hope that, in this way, both logic and history will be satisfied.

We shall therefore find again 4 main sections, of which two embrace the solely **Rate**-Controlled SCTA techniques (including those previously called Controlled Rate Thermal Analysis (CRTA) and Quasi-Isothermal Techniques) whereas the two others embrace the SCTA techniques which are controlled both by the *Rate* and by the *Temperature* (including the Stepwise Isothermal Analysis (SIA) and some complex forms of SCTA). The chapter ends with a technique closely related with SCTA in spite of being, strictly speaking, out of its scope.

3.2. Constant Rate Thermal Analysis

This was the first form of SCTA ever developed, probably because (i) it is relatively straightforward and (ii) it allows, when a low rate of transformation is selected, to fulfill the requirements of high homogeneity and high resolution which were initially looked for.

It is the simplest form of thermal analysis based on Figure 2-3.

It must be pointed out that all techniques described in this section require that the preselected rate of transformation be reached at the start of the Constant Rate experiment proper. This start may require an initial heating of the sample, which is simply obtained with a temperature ramp. Most often, nothing measurable happens in the sample during this preliminary period which is therefore considered to be out of the scope of the thermal analysis proper.

3.2.1. CONSTANT HEAT-FLOW THERMAL ANALYSIS OR CONSTANT RATE DTA

"*Simple Method of Thermal Analysis*" (C.S. Smith, Waterbury, Connecticut, 1940)

The aim of this experiment was to permit "quantitative measurements of specific and latent heats" in metallic alloys, "in the rather limited field where reactions are fast enough" (i.e. are "completed at rates of temperature change above about 1°C per minute") [1]. In other words, it aims to be a quantitative DTA to study phase changes in alloys.

The original sketch of the experiment is given in Figure 3-1, with a 2-inches scale in it. The sample (in black, in the center) fits closely in a refractory container, with low thermal conductivity, surrounded by a furnace (nichrome winding on a thin alundum core). Two thermocouple junctions (cf Fig. 3-1,b) allow to measure the temperature difference between the inner and outer wall of the ceramic container. This temperature difference is kept constant by appropriate heating of the furnace, so that the heat flow through the ceramic wall is approximately constant (it varies, however, because of the change of thermal conductivity with temperature). The system can be operated on heating or cooling. Sample masses used in this paper are 86 and 150 g (brass or copper alloy) and the temperature difference maintained through the ceramic wall is of the order of 60 K. It is claimed that this new method lowers the temperature gradients within the specimen and produces sharper peaks (corresponding to phase changes) on the Reverse of the cooling rate *vs.* Temperature recording.

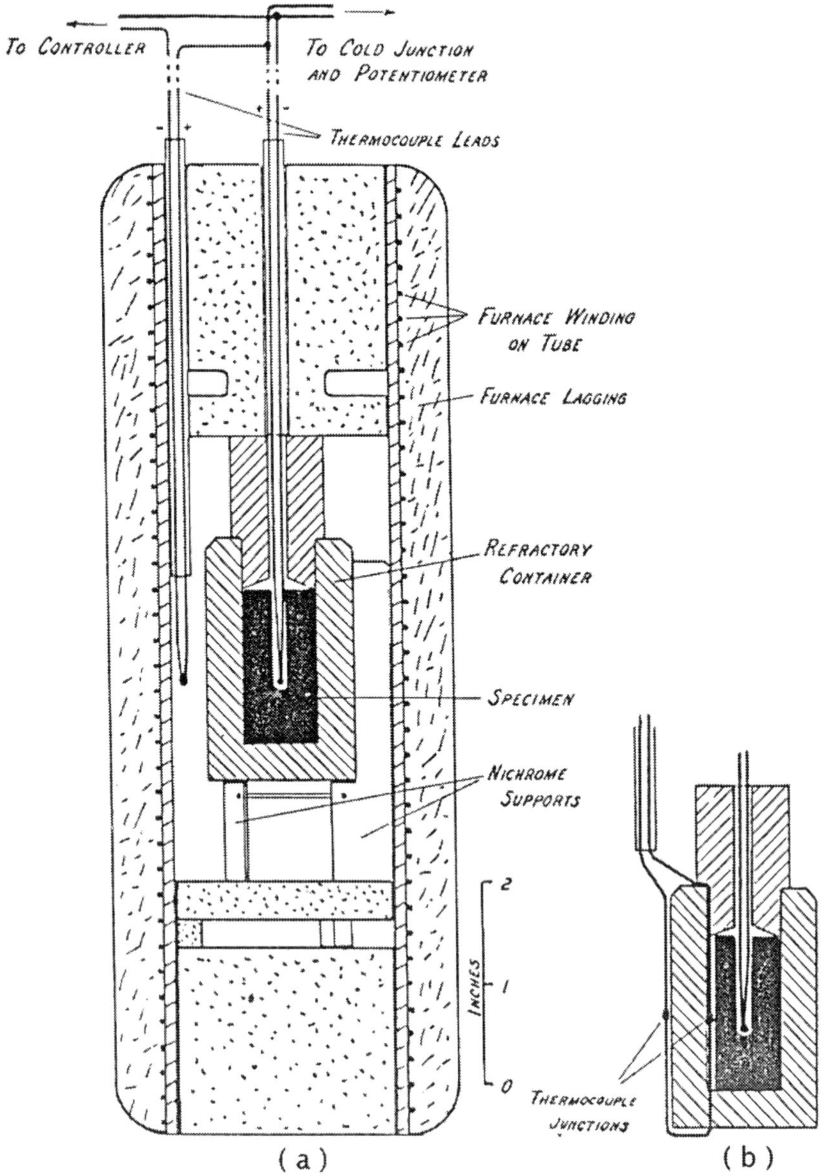

Figure 3-1. C.S. Smith's apparatus (after [1]).

"*Thermal Analysis Equipment with Constant Heat Flow*" (R. Cohen-Adad and B. Boinon, Lyon, 1974)

This equipment was designed to determine phase-diagrammes of salts (fluorides, to begin with) between –100 and +180°C [2]. It makes use of the same principle as Smith, but with a rather different design: the temperature drop is maintained constant through an air-sleeve (not through a ceramic wall), a second, exterior, furnace (also made of a thin cylinder of brass), is kept at the same temperature as the main one (in order to provide an adiabatic shield and increase the stability and sensitivity), provision is made for cooling by liquid air and, finally, the whole system can be rocked (3 seconds per cycle) to increase the homogeneity of the sample on melting. It is claimed that, contrary to conventional thermal analysis, the length of the isothermal plateaus (which are recorded when a first order transition takes place) is here directly proportional to the enthalpy of phase change, whereas the separation power of successive thermal events is increased.

"*Quasi-Isothermal DTA Method*" (J. Paulik, F. Paulik and M. Arnold, Budapest, 1985)

This experiment, which is mounted on a complete "Derivatograph", now makes use of a true differential mounting (with sample and reference) contrary to the two former (which did not use a reference) and needs, of course, much smaller samples. The physical quantity acting on the furnace heating (cf Fig. 2-3) is directly the DTA signal. Like most experiments in this section, it may require an initial temperature ramp (with constant heating rate) and it switches to a real Controlled Rate mode as soon as the onset of the thermal reaction is detected by the DTA sensor. Now, a feature of this experiment is that it actually makes use of the first derivative of the DTA signal with respect to time to detect this onset: this is an elegant way not to be annoyed by the often important shift of the DTA base-line, which makes it difficult to decide in advance of the value of a threshold which would operate the switch to the Controlled Rate mode. The shift of the first derivative is much smaller, yet with a high sensitivity.

3.2.2. CONSTANT RATE EVOLVED GAS DETECTION (CR-EGD)

"*Thermal Analysis under Low Pressure and Constant Decomposition Rate*" (J. Rouquerol, Paris, 1964) and "*Constant Rate Thermal Analysis*" (J. Rouquerol, Marseille, 1970)

Figure 3-2 provides us with a simple representation of the most basic equipment of Constant Rate Evolved Gas Detection (CR-EGD) used by

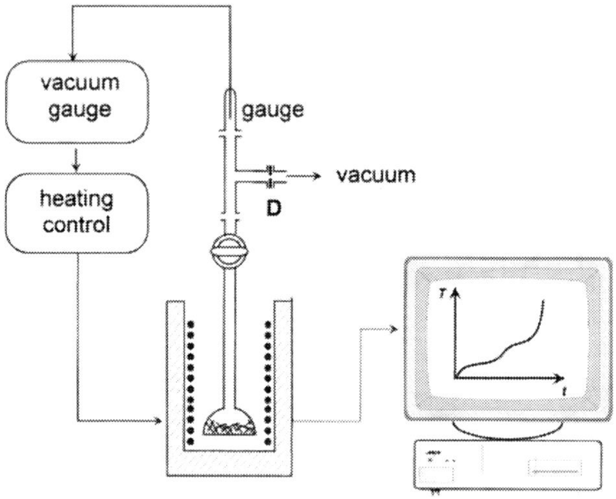

Figure 3-2. Schematic representation of a basic CR-EGD equipment.

Rouquerol *et al.* since the early sixties. The sample, visible here in the bottom of a glass bulb, is permanently evacuated through the diaphragm "D", which is simply a constriction of any type. The residual pressure above the sample is monitored by a vacuum or pressure gauge (Pirani gauge, Penning gauge, capacitive pressure transducer …) whose electrical output is sent (in place of the usual temperature sensor output) to the heating controller. The latter heats the furnace in such a way as to keep the residual pressure at a constant, preset, value.

Controlling this residual pressure has a number of interesting effects:

1) The reaction rate can be kept constant. This is true when the gas evolved has a constant composition, in particular when it is a pure gas (H_2O, CO_2, etc.): the constant pressure difference between both sides of the diaphragm produces a constant gas flow, which is itself exclusively due to the gas evolving from the sample, therefore exactly at the same rate. This set-up can therefore provide, in particular, all advantages expected from the selection of a very low reaction rate, i.e. lowering at will of the temperature and pressure gradients which are a source of heterogeneity within the sample and avoiding any "spurting out" of the sample during a thermal decomposition under low pressure

2) The simple recording of the sample temperature *vs.* time provides a full thermal analysis curve, directly comparable to a TG curve, in spite of the much simpler and tougher set-up: because of the constant gas flow, the mass lost by the sample is indeed directly proportional to time.

3) The residual pressure not only above but also within the sample (specially when it is a powder), provided the gradients are kept low enough, is also kept constant. We shall see for instance in Chapter 6 and 7 how important this parameter can be for the preparation of adsorbents and catalysts.

4) This residual pressure control is extremely sensitive and provides, in practice, an illimited sensitivity (keeping for instant constant, if needed, a rate of CO_2 loss as low as 1 mg *per day*!)

A major feature of this equipment is that the sample is located in a pyrex glass or fused silica bulb, equipped with a stopcock, which is directly suited for further connection to a standard manometric gas adsorption equipment. This set-up is therefore specially suited for the study of textural changes (i.e. surface area and pore-size distribution) of a material during its thermolysis and, moreover, in extremely well-controlled conditions [6]. It must also be stressed that the thermolysis can be stopped at any interesting point of the thermal analysis curve, allowing the sample to be characterized by gas physisorption before being connected again to the CR-EGD set-up to carry out a further step of the thermolysis. The sample remains within the same, standard "adsorption bulb".

This set-up is also well-suited for the thermal preparation of powders or porous solids which are to be studied by immersion calorimetry, a technique which is very sensitive to the surface state and to the outgassing conditions: the sample bulb is now equipped with a brittle end; at the end of the thermal treatment the bulb is sealed (therefore kept under vacuum) and then introduced into the calorimeter before beeing broken [6].

The pressure range in which this type of set-up was operated lies between 10^{-3} and 20 mbar, which happen to be the most interesting for the production of adsorbents from inorganic precursors. The usual temperature range is between –30 and +1000°C.

Figure 3.3 (a) shows an all-glass set-up used in the early sixties. The heating control, which is critical in this type of experiment, was made proportional by addition of a modulation to the Pirani gauge signal and made integral by use of a motor-driven variable transformer [7]; there was provision for several possible pumping rates, by using one or several of the three stopcocks giving access to the vacuum manifold. Figure 3.3 (b) represents a more recent, all-metal, set-up whose design is close to those used at present. The stability of the Pirani gauge signal was here improved by use of a small thermostat around it, kept at a constant temperature (between 30 and 40°C), within 0.1 K. The diaphragm "d" was a metallic, interchangeable, disk with a hole (diameter between 0.4 and 5 mm). Other details can be found in [8]. Today, a turbo-molecular pump is also a good choice (specially when there is any risk of catalytic decomposition of oil

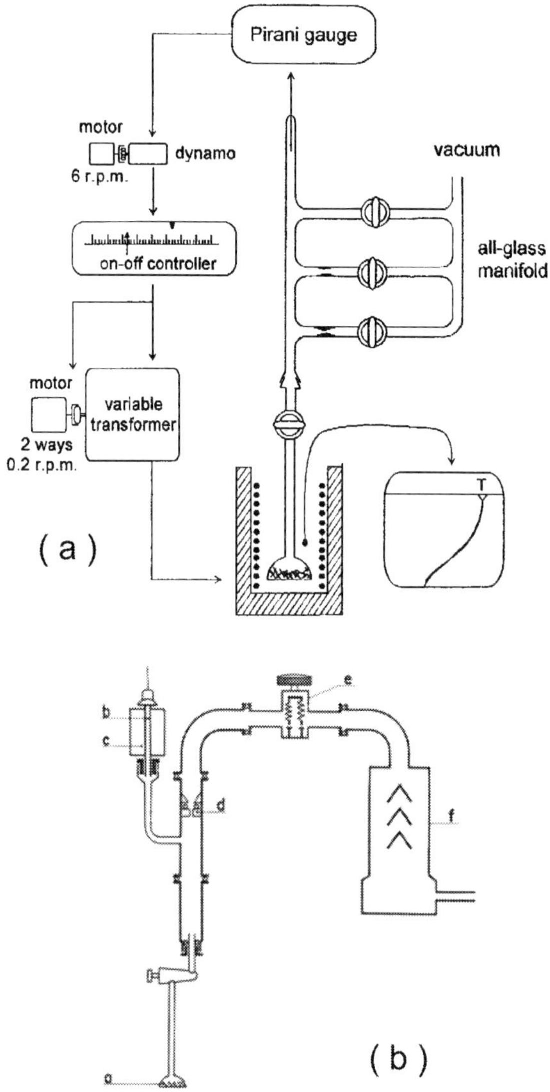

Figure 3-3. Two forms of CR-EGD equipment used by J. Rouquerol: (a) all-glass (early sixties (7)) and in 1962–1966 (7) and (b) all-metal (late sixties (8) a: sample bulb, d: diaphragm, e: bellow-valve, f: diffusion-pump, b: Pirani gauge, c: thermostat around the Pirani gauge.

vapour on the sample) and the stability of some modern Pirani gauges does not justify any more their being protected by an extra thermostat. Any type of vacuum gauge can be used, provided it delivers a stable electrical signal and it is able to respond in a few seconds: Pirani, Penning, membrane gauge (using, as the sensor, either a capacitance, or a strain bridge ...). At first sight, a set-up like that of Figure 3.3 (b) does not allow to independently study the influence of the rate of transformation and that of the residual pressure on the course of the thermolysis: as soon as you change the setting of the residual pressure you get a different gas flow rate through the diaphragm. There is fortunately an easy way to overcome this apparent impossibility. In case, indeed, one is interested *to study the influence of the pressure alone* (but keeping the rate of transformation constant) one simply has to use a cascade of two successive diaphragms and two successive pressure (or "vacuum") gauges. The second gauge and diaphragm (those situated downstream) are those used to control the experiment and to provide a constant downstream pressure (which is not the pressure surrounding the sample) and a constant rate of gas evolution. From one experiment to the other, one does not change these settings, but simply changes the opening of the first diaphragm (upstream): this modifies at will the residual pressure around the sample. It is in this way that the curves of Figures 3, 5 and 7 (b), in Chapter 6, which show the influence of the residual water vapour pressure on the thermal decomposition of gibbsite, were obtained. Now, in case one wishes *to study the influence of the rate of transformation alone* (but keeping the residual pressure constant), which can be useful for kinetical studies, one simply has to use the arrangement of Figure 3.3 (b), where only the diaphragm is changed from one experiment to the other. Even the set-up of Figure 3.3 (a) is suitable for that and was effectively used with that procedure [9].

A modification was proposed by Fierro *et al.* in 1990 [10], which is to use, for the heating control, not one but two Pirani gauges, with the second one on the vacuum side of the diaphragm, so as to take into account, if needed, any change in the vacuum provided by the pumps. It has not been our experience that this association of two Pirani gauges was ever necessary: we prefer to check that the vacuum provided by the pumps is at least 2 orders of magnitude better than the residual pressure desired over the sample, which is the only way to ensure, during the experiment, both a constant pressure and a constant rate.

Simultaneous CR-EGD and Thermogravimetry or Calorimetry

Figure 3-4 shows 2 ways for extending the possibilities of Controlled Rate Evolved Gas Detection, by carrying out simultaneous experiments on a single sample.

Figure 3-4 (a) represents, with Loop II, the association of CR-EGD *with thermogravimetry* in case a vacuum balance is available: the system operates in the same way as the one described in the preceding section, except that, at the same time, the sample mass is recorded. This was the principle followed in 1968 by Rouquerol to transform in one day a commercial vacuum thermobalance into a Constant Rate Thermal Analysis equipment, by simply using the vacuum gauge and the good quality heating control already available on the thermobalance [13]. This system was specially suited to control very low residual pressures: it operated essentially in the 10^{-5}–10^{-3} mbar range, in which a high constancy of this pressure (usually called vacuum …) was achieved.

Figure 3-4 (a) represents, with Loop II, the association of CR-EGD *with thermogravimetry* in case a vacuum balance is available: the system operates in the same way as the one described in the preceding section, except that, at the same time, the sample mass is recorded. This was the principle followed in 1968 by Rouquerol to transform in one day a commercial vacuum thermobalance into a Constant Rate Thermal Analysis equipment, by simply using the vacuum gauge and the good quality heating control already available on the thermobalance [13]. This system was specially suited to control very low residual pressures: it operated essentially in the 10^{-5}–10^{-3} mbar range, in which a high constancy of this pressure (usually called vacuum …) was achieved.

Figure 3-4 (b) represents the association of CR-EGD *with calorimetry,* as it was developed in 1971 by Ganteaume and Rouquerol [12]. The figure looks even more similar to Figure 3-2, since the furnace seems to be simply and directly replaced by a calorimeter. The reality was somewhat complicated, because of the need of microcalorimetry (since the thermal effects are spread over periods of the order of one or two days, which provides a small signal)

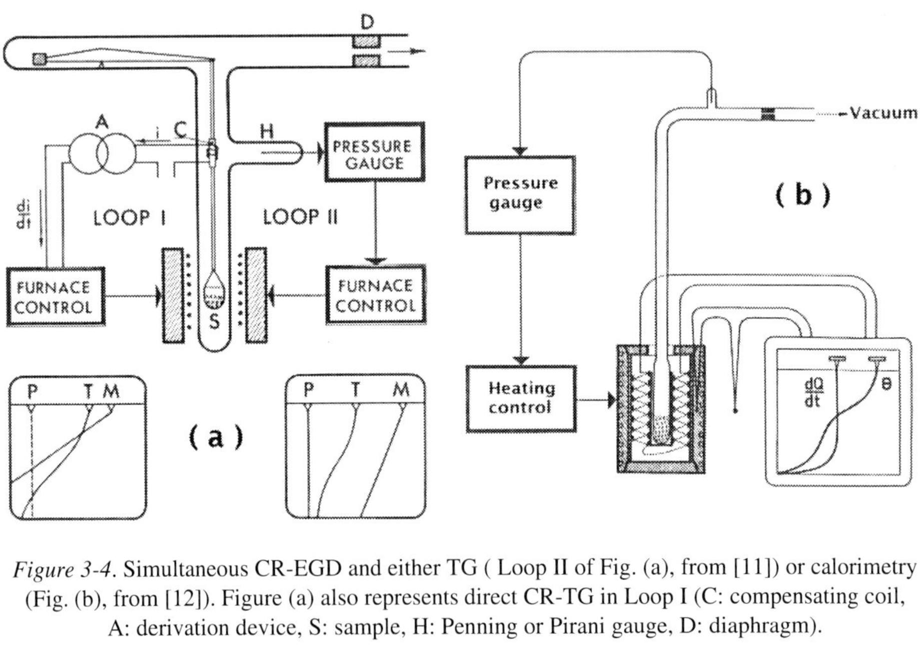

Figure 3-4. Simultaneous CR-EGD and either TG (Loop II of Fig. (a), from [11]) or calorimetry (Fig. (b), from [12]). Figure (a) also represents direct CR-TG in Loop I (C: compensating coil, A: derivation device, S: sample, H: Penning or Pirani gauge, D: diaphragm).

and because of the extreme sensitivity of microcalorimetry to any thermal perturbation brought by the heating controller. A special differential microcalorimeter (based upon Tian-Calvet thermopiles) had to be built for this experiment, with a double thermostat and a short response time for the inner one, in which, in spite of the sample-controlled heating, the temperature fluctuations were smaller than 5×10^{-3} K. This set up was used to measure the enthalpy of transformation of gibbsite into alumina, in well controlled conditions.

Constant Rate Evolved Gas Detection or Analysis Under a Flow of Inert or Reactive Gas

Here, a gas (inert or reactive, under atmospheric pressure) passes over the sample and its final composition is analysed. In this experiment, the final composition is the physical quantity which is kept constant through an appropriate heating of the furnace. This can be considered as an experiment of Evolved Gas Analysis (EGA) or of Reacted Gas Analysis, depending on the inert or reactive nature of the gas and, from this viewpoint, can then be called, more precisely "*Constant Rate Evolved (or Reacted) Gas Analysis under flow of inert or reactive gas*".

The first experiment of this type was set up by Stacey in 1981 (use of a helium flow and of two detectors, a katharometer and a dew-point hygrometer [14]; the same assembly was also used with a reactive gas – hydrogen – to study the reduction of catalysts [15].

M. Reading [16] used two infra-red detectors (cf Fig. 3.5) to monitor, independently, CO_2 and H_2O, in order to follow the thermolysis of transition metal hydroxy carbonates, using N_2 as the vector gas. With a multi-reflection gas cell, he could control the heating of the sample from the partial pressure of CO_2 (which was kept constant around 2 mbar) and could then simply monitor the corresponding partial pressure of H_2O and therefore compare the rate of evolution of the two gases. Arellano, Criado and Real used a catharometer and, with either helium or hydrogen, followed the thermal decomposition of $CaCO_3$ and $Ca(OH)_2$ and the reduction of NiO [17]. With a similar equipment, they could also carry oxidation studies [18].

Barnes, Parkes, Brown and Charsley [19] used either a hygrometer or a katharometer or a mass spectrometer to analyse the gas flowing over the sample.

Rouquerol and Fulconis [20] used a very sensitive differential gas flowmeter based on the principle of a pneumatic Wheatsone bridge in order to detect any gas evolution (in an inert carrier gas) or consumption (of a reactive gas) [20].

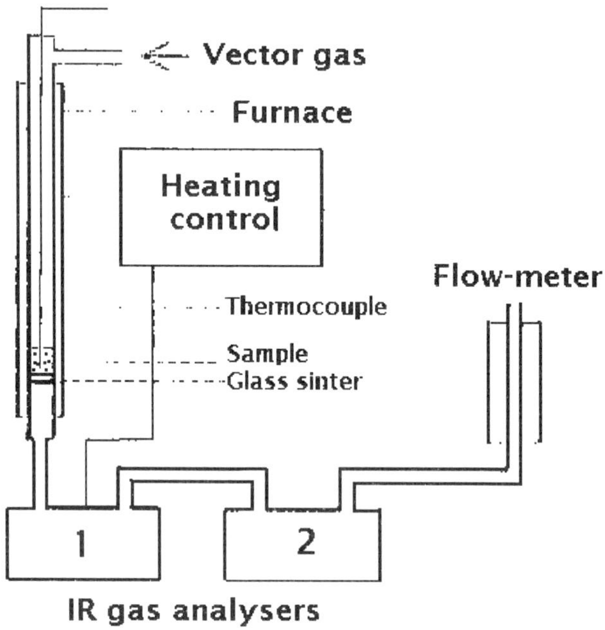

Figure 3-5. CR-EGA set-up with vector gas and two IR analysers (after [16]).

3.2.3. CONSTANT RATE THERMOGRAVIMETRY (CR-TG)

Quasi-Isothermal Quasi-Isobaric TG (Paulik, Paulik and Erdey, Budapest, 1965–1971)

The name "Quasi-Isothermal Quasi-Isobaric TG" was proposed by J. and F. Paulik in 1971, when they published their first results obtained with this method [21], but the "quasi-isothermal" principle was already described in their first patent published in 1965 [22], where a figure similar to Figure 3-6 can already be found.

The DTG signal (directly obtained by electromagnetic derivation, by help of a coil, attached to a moving arm of the balance, and a magnet) is sent to the heating control of the furnace which, basically, acts in such a way as to keep this signal approximately constant, for instance thanks to an on-off control of the voltage available to heat the furnace [23]. This type of experiment was less easy to get working than the CR-EGD experiment described in Section 3.2.2, simply because the stability and sensitivity required from the DTG signal in order to really get a slow rate of transformation was extremely demanding for the technical possibilities of the sixties. This is what J. and F. Paulik explain page 48 of their book [23]: "We made the first attempt in 1962 to develop this technique, but it was only in 1971 that every obstacle to the realization of an appropriate measuring system could be eliminated". It is also in 1971, with the first published experiments, that the "labyrinth crucible" was proposed and used [21]: the furnace [23]. This type of experiment was less easy to get working than the CR-EGD experiment described in Section 3.2.2, simply because the stability and sensitivity required from the DTG signal in order to really get a slow rate of transformation was extremely demanding for the technical possibilities of the sixties. This is what J. and F. Paulik explain page 48 of their book [23]: "We made the first attempt in 1962 to develop this technique, but it was only in 1971 that every obstacle to the realization of an appropriate measuring system could be eliminated". It is also in 1971, with the first published experiments, that the "labyrinth crucible" was proposed and used [21]: the crucible and cover are shaped in such a way as to offer a long, compulsory, path to the escaping gas; in these conditions, as soon as the gas produced by the thermolysis is able to fill and "wash" the crucible by pushing the air out, the sample is completely surrounded by 1 bar of self-generated gas. The term "quasi-isobaric" therefore does not only mean that the experiment is carried out under 1 bar (which would simply require not to use a tight crucible cover) but it means "under I bar of self-generated atmosphere".

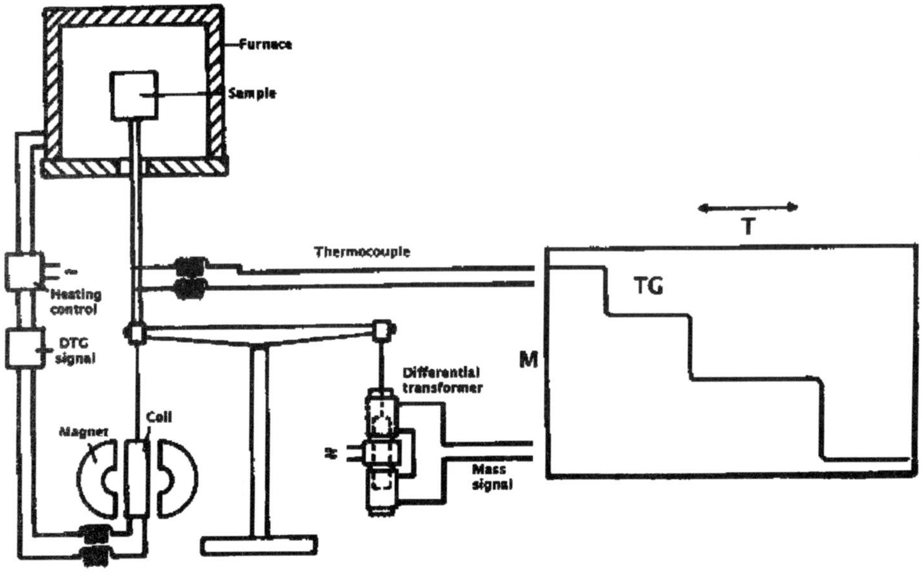

Figure 3-6. Principle of quasi-isothermal TG (after [23]).

Quasi-isothermal quasi-isobaric TG proved to be a most fruitful technique, complementing well the Constant Rate EGD which was already running: the latter indeed is specially interesting for studies under low pressure (where this experiment can reach extremely high sensitivities) whereas the former is well suited for studies under atmospheric pressure. A most comprehensive review of all work obtained until 1995 by his approach can be found in F. Paulik's book [24].

3.2.4. CONSTANT RATE THERMODILATOMETRY (CR-TD)

Rate Controlled Sintering (Palmour and Johnson, 1967)

Palmour and Johnson proposed this experiment with the aim of improving the final quality of the ceramic, thinking that this would be favoured by the control of the rate of sintering [25]. For this purpose, they used an appropriate connection between the signal of a standard thermodilatometer and the heating control of its furnace. These authors do not seem to have continued in that direction and did not publish any more on this topic.

Quasi-Isothermal Thermodilation (Paulik and Paulik, 1977)

The spirit is that of Palmour and Johnson's "Rate controlled sintering", where one wishes to obtain an approximately constant rate of shrinkage, but the experiment was cleverly developed [26] through a relatively minor modification of the quasi-isothermal thermobalance described in Section 3.2.3: the sample is made in the form of a tablet and placed on the flat pan of the balance; its top is in contact with a fixed quartz rod which prevents it from moving upwards. In case of dilatation, the moving arm of the balance is depressed. The corresponding displacement is recorded (by means of the differential transformer visible in Fig. 3.6) whereas a signal proportional to its derivative *vs.* time is directly obtained by the coil and magnet device and sent to the heating control. A few examples of application of this technique can be found, again in [23] and [24].

3.3. Varying Rate Thermal Analysis

3.3.1. RATE-JUMP EGD AND RATE-JUMP TG

The "Cyclic Heating" or "Rate-Jump" Method (J. and F. Rouquerol, 1971–1985)

Until 1971, all what we would call to-day "SCTA experiments" were carried out with the objective of keeping the rate of the transformation as constant as possible. It is for kinetic purposes that a procedure involving a varying rate was actually introduced [27]. Its principle was to periodically switch from rate of transformation r_1 to rate of transformation r_2 and to record the corresponding sample temperatures T_1 and T_2, as represented in the left part of Figure 3-7. The “asumptionless” determination of the energy of activation (i.e. without any asumption about the mechanism involved) which can then be independently carried out during each cycle is explained in detail in Chapter 4.

Coming now to the basic techniques available to achieve this result, we can say that it is specially easy and precise *with a relatively standard equipment of CR-EGD*: one only needs, to get the highest rate r_2 to by-pass the first diaphragm by a second one. An easy way to get exactly the ratio of rates desired, whatever the pressure and the gas flow regime, is to simply use, for the two diaphragms, two identical metal disks, for instance 1 cm thick, and to use the same drill to make one hole in the first disk and, say, 3 or 4 holes in the second one: no extra measurement is needed to know that the pumping rates will be in the ratio of 3 or 4. This is also easy to automate, by simply using an electrovalve to govern the by-pass and by controlling it by a computer or by a simple timer. After calling it the “Cyclic heating” method [27] the authors call it, since 1985, the “Rate-jump method” [29].

This method was used early by *associating CR-EGD with TG* [30]: in this case, CR-EGD ensures the control of the transformation rates, whereas TG is used to measure the rates from the successive slopes of the mass *vs.* time curve, as illustrated in Figure 3-8. This allows to replace the two diaphragms mentioned above by any type of non-calibrated valve. This was found specially interesting when the measurements have to be carried out under a pressure (or, better said, a vacuum...) ranging between $2\times10^{-5}$and 10^{-3} mbar, where vacuum valves with large openings are requested. For instance, the set-up represented in Figure 3-8 used a butterfly valve B (compressed-air operated) with 5 cm bore, which proved to be necessary to maintain the desired vacuum around a 100 mg sample of calcium carbonate, during its decomposition. Incidentally, any other approach than SCTA would inevitably “break the vacuum” as soon as the decomposition temperature is reached and could therefore hardly provide any reliable information in this pressure range.

Another way to achieve rate-jump measurements under a high vacuum is of course to use CR-EGA with a mass spectrometer as the detector. This was used to determine, for the first time, the activation energy of the production of one single component (here, CO2) of the gas mixture evolved during a complex thermolysis (here, that of kerogens) [30]. It is with a similar equipment that the experimental results presented in the right part of Figure 3-7 were obtained. This figure illustrates the large number of independent, automatic, determinations of the activation energy which can be carried all along the thermolysis of a single sample: here *ca* 100 successive measurements on one sample of hexahydrated uranyl nitrate [28]. Let us also stress that very nice rate-jump measurements of the energy of activation which were carried out by Tiernan, Barnes and Parkes on samples smaller than 10 mg, thanks to the use of solid insertion probe mass spectrometry [31].

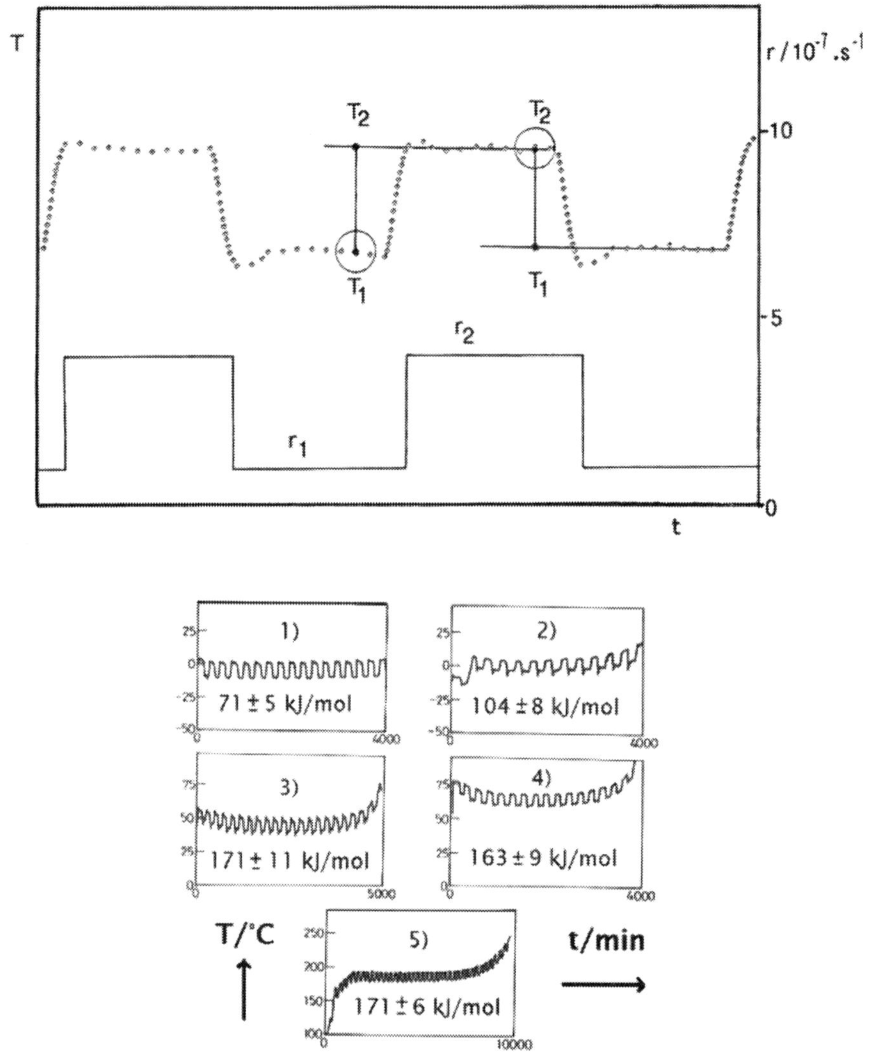

Figure 3-7. Principle of the Rate-Jump method (left) and results obtained over the 5 steps of the thermolysis of hexahydrated uranyl nitrate (steps 1–4: dehydration; step 5: denitration), after [28].

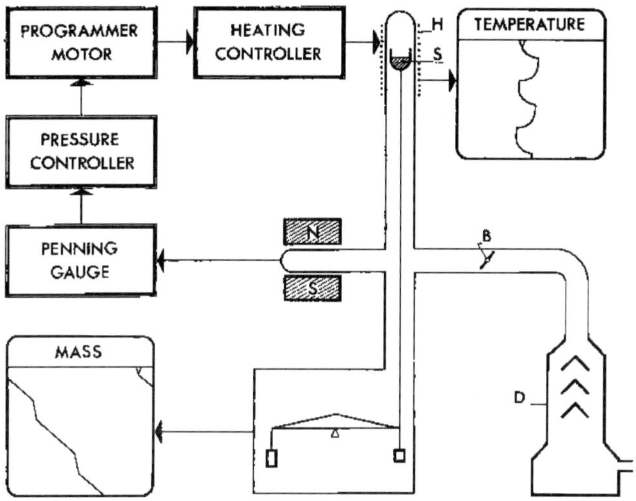

Figure 3-8. Association of CR-EGD and TG for the application of the Rate-Jump method under pressures down to 2×10^{-5} mbar (from [30]).

3.3.2. CONTROLLED RATE EVOLVED GAS ANALYSIS (CR-EGA)

Controlled Decomposition Rate Evolved Gas Analysis (Thevand, Rouquerol and Rouquerol, 1982)

The word "Controlled" was proposed in 1982 instead of "Constant" to embrace the cases where *a varying composition of the evolved gas* still allows to control its rate of evolution but not, strictly speaking, to keep the rate of decomposition constant [30]; one can consider it as a kind of loosely controlled CRTA experiment.

In practice, once the thermolysis is known to be complex, one must analyse the gas phase if the wishes to understand anything in the mechanism. A quadrupole mass analyser is among the most suited and versatile detectors which can be used in this situation. Nevertheless, after several unsuccessful trials in the preceding years, it is only in 1982 that the stability of the signal provided by quadrupole mass analysers began to fulfil the requirements of the demanding control loop of a SCTA experiment and allowed the publication of the first results [30]. Figure 3-9 provides us with two representations of a Controlled Rate EGA equipment. The figure in the left illustrates a major interest of this experiment: of the three partial pressures shown to be recorded on the chart, one is constant. The latter partial pressure is indeed that of the gas which was chosen to control this sample-controlled experiment. The CR-EGA set-up is therefore heating the sample in such a way as to produce this gas at a constant rate, irrespective of what happens to the other gases produced by the

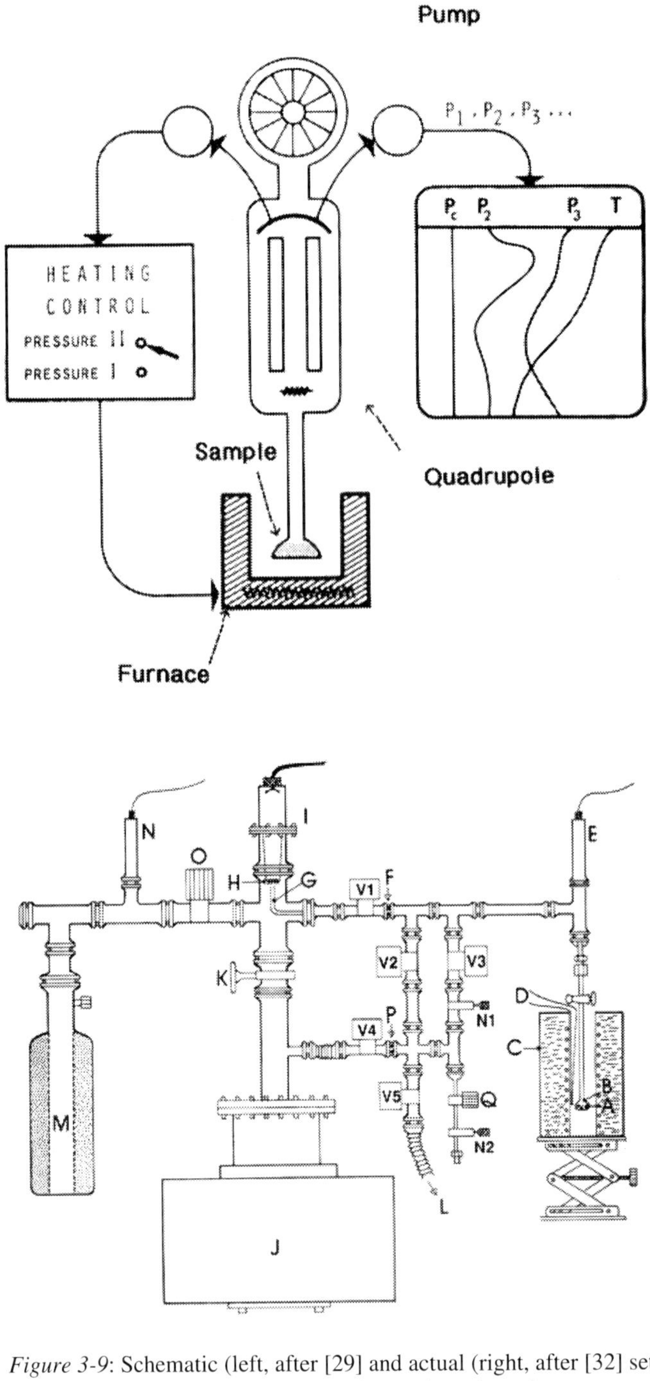

Figure 3-9: Schematic (left, after [29] and actual (right, after [32] set-up of CR-EGA making use of a quadrupole analyser.

complex reaction. For instance, in a thermolysis producing all at once CO_2 and H_2O, it was possible to carry out distinct experiments where either CO_2 or H_2O was the controlling gas (and, incidentally, to measure the energy of activation of the production of that single gas, as reported to in the preceding section) [30].

The right part of Figure 3-9 shows an actual experiment, when one sees the sample (A) in its bulb (B), a Pirani gauge (E), the quadrupole analyser "I" with its heated filament "H", the bent tube "G" (able to conduct just underneath the filament, when needed, the whole of the gas evolving from the sample), the ionic pump "J" and the molecular sieve cryogenic pump "M" used to start the ionic pump. Let us point out a few features about this experiment:

- Three distinct signals can operate the heating control of furnace "C" in the SCTA mode, i.e. either the signal from the Pirani gauge "E", or the partial pressure of any gas detected by the quadrupole analyser (as commented on above) or, still, the signal provided by the ionic pump, since the current through that pump is directly related to the rate of pumping.
- Because of the possible control of very low rates of gas evolution from the sample, it is possible to send the whole gas produced through the quadrupole analyser, thus avoiding any of the problems of discrimination (between small and large molecules) inevitably introduced by all sampling systems making use of a leak.
- Valves V2 and V4 allow to by-pass diaphragm "F" by diaphragm "P": these valves were used to provide the Rate-Jump experiments shown in the right of Figure 3.7.
- Incidentally, this equipment happens to be totally silent and does not need any water supply: it happened to be installed and to be used, during a while, in a library ....

3.3.3. INCREASING RATE THERMAL ANALYSIS

Increasing Rate Thermal Analysis (Ortega, Pérez-Maquéda and Criado, 1994)

This procedure was suggested by A. Ortega, L.A. Pérez-Maquéda and J.M. Criado for kinetic purposes which are developed in Chapter 4 [33]. The aim was to increase the discrimination power of SCTA with regard to the mechanism involved. These authors have shown that *increasing the rate of transformation linearly* could bring, in some cases, the missing information. It is worth noticing that such an "exotic" way of programming the SCTA experiment is not specially difficult and can be done with standard programmers for heat control, keeping in mind that programmer is now fed by the signal related to the rate of transformation, instead of the usual temperature signal.

3.3.4. MODULATED RATE EGD AND MODULATED RATE TG

Modulated Rate EGD (J. Rouquerol and J.M. Fulconis, 1996)

One easily "jumps" from the idea of a Rate-Jump procedure to that of a modulation. Now, the relatively long heating control loop of a SCTA experiments does not made it very easy to get a modulation. Nevertheless the idea of these authors was of simply taking advantage of the natural tendency of any SCTA experiment to oscillate around the preset value of the transformation rate, specially when the heating control is coarse and close to an on-off control (20). Provided on has a means to simultaneously record the rate of transformation (for instance, with help of an appropriate gas flowmeter or with a thermobalance) as the same time as the temperature, one then can directly draw the activation energy (like with the rate-jump method) simply from the amplitude of the oscillations of these two signals.

Modulated TG (R. Blaine, 1997)

Introducing a modulation in a TG experiment was in line with the somewhat earlier start of modulated DSC [36]. Now, R. Blaine examined and developed it in two ways, i.e. by imposing either a temperature modulation or a rate of weight loss modulation [34,35]. The first mode is in the field of temperature-controlled thermal analysis (or conventional thermal analysis) whereas the second mode is in the field of SCTA, as represented in Figures 2.3 and 2.4 in Chapter 2. To be used in kinetics, this approach is demanding from the technical viewpoint, but it looks promising.

3.4. Alternate Rate- and Temperature-Controlled Thermal Analysis

A characteristic feature of this technique is that the transformation rate of the sample is continuosly monitored during the heating between the transformations, which is carried out at constant heating rate as in conventional TA. As soon as the transformation rate exeeds a preset limit the controlling conditions are immidiately changed to the conditions needed for the measurement being carried out, this can be constant temperature as in SIA (Stepwise Isothermal Anlysis) (Sorensen [37]) or heating at a lower rate as in the Proportional Heating technique introduced by Parkes, Barnes and Charsley [38]. This technique can therefore be considered as a true sample controlled technique as both the advent of the transformation as well as the conditions during the transformation is controlled by a feed back from the sample itself.

Among the techniques based on this principle can be mentioned:

– Stepwise Isothermal Analysis (SIA) (Sorensen)
– Forced Stepwise Isothermal Ansalysis (FSIA) (Sorensen)
– Proportional Heating (PH) (Parkes, Barnes and Charsley)
– Maximum Resolution (Max Res) (Schenker and Riesen)
– Peak Slope Heating (Parkes, Barnes and Charsley)
– Dynamic Heating (Opferman)

3.4.1. STEPWISE ISOTHERMAL ANALYSIS (SIA) (SORENSEN, 1978)

Stepwise Isothermal Analysis (SIA) introduced by Sørensen [37] in the late seventies, and which can be used both in thermogravimetric and dilatometric measurements (which were initially called by the author "Quasi-Isothermal techniques"), is a typical "event-controlled" technique, i.e. when an "event" such as a thermal decomposition, an evaporation, a reduction or an oxidation or a shrinkage during sintering is starting in the sample during the continuous heating as performed in conventional thermal analysis, then a change is triggered in the mode controlling the measuring conditions. As shown in Figure 3-10 the overall controlling parameter in SIA and related techniques is the derivative weight (dw/dt) or lenght (dL/dt) signal, corresponding to the reaction rate, which is calculated in the algoritm after each weight (length) measurement. As long as this signal is smaller than a preselected treshold value, the heating is allowed to proceed at the preset constant heating rate, but the moment this signal becomes larger than the threshold value, then the heating is stopped and the reaction takes place at isothermal conditions. After completion of the reaction the heating is resumed when the derivative signal again becomes smaller than the treshold value and the sample will be heated at constant heating rate until the next reaction temperature is reached. The overall measurement will therefore take place in characteristic isothermal steps depending of the number of reactions for the sample as illustrated in Figure 3-11, which shows the SIA curves obtained in a study of the thermal decomposition of Ce-carbonate ($Ce_2O(CO_3)_2H_2O$) in different atmospheres [39].

An important feature of this technique is thus that it provides isothermal data which are suitable for a determination of the reaction mechanism in kinetic studies. A drawback, however, is that it only provides data at constant temperature and it cannot therefore be used to determine the activation energies. In order to overcome this difficulty the technique Forced Stepwise Thermal Analysis (FSIA), which is described in the next section, was introduced.

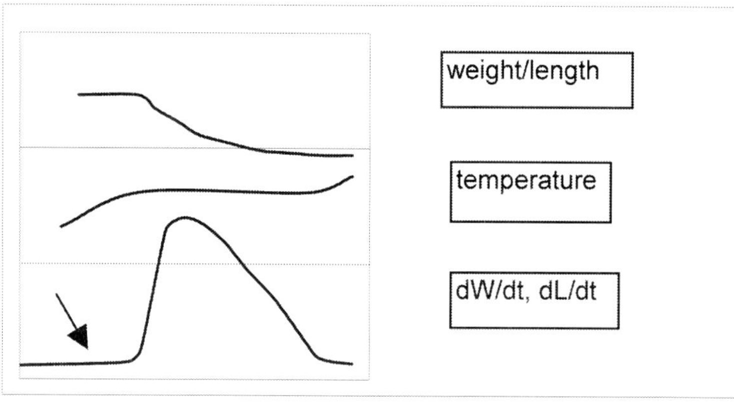

Figure 3-10. Principle of SIA. Arrows indicate threshold values at the start and the end of the reaction. The threshold value at the end is smaller that that used at the start.

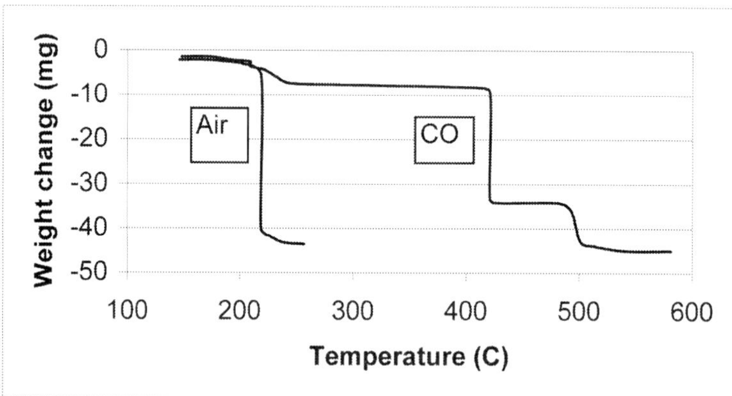

Figure 3-11. SIA for the thermal decomposition of Ce-carbonate in air and CO.

As also shown in Figure 3-10, two different threshold values are generally applied in SIA: one value, which is the highest, for detecting the onset of the reaction and a much smaller value at the end of the reaction. In this way the "tail" of the reaction, which in many cases is very long, can also be taken into account and which considerably improves the resolution. This was demonstrated by Sorensen in a study of the thermal decomposition of copper sulphate pentahydrate ($CuSO_4 \cdot 5H_2O$) (Fig. 3-12), which showed that the release of water could be resolved into much finer steps than the overlapping steps generally obtained in conventional thermal analysis measurements.

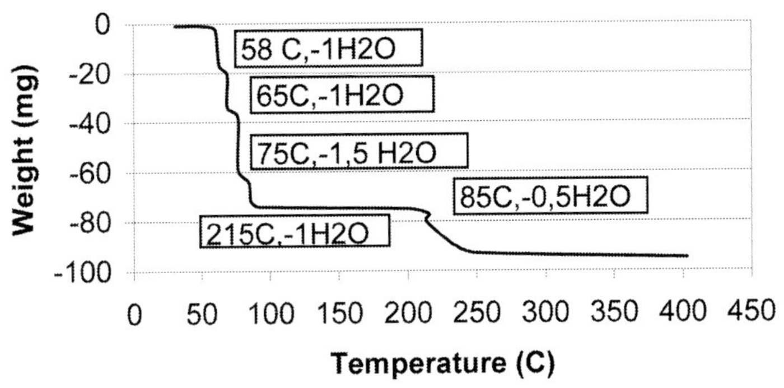

Figure 3-12. High resolution obtained by SIA for the thermal decomposition of copper sulphate pentahydrate.

3.4.2. FORCED STEPWISE ISOTHERMAL ANALYSIS (FSIA) (SORENSEN, 1992)

The drawback of SIA, that it only provides data at constant temperature, was overcome by introducing FSIA which operates in the following way: when the weight change, which is constantly monitored during the reaction, reaches 10% of the total weight change for a particular relation, the temperature is forced to increase with 5°. The total reaction is in this way forced to take place in about 10 isothermal steps at increasing temperature as illustrated in Figure 3-13, which shows the FSIA curves obtained in a study of Ce-carbonate in different atmospheres [39]. That this technique is suitable for the determination of the activation energies for these reactions is demonstrated in Chapter 5.

3.4.3. STEPWISE TEMPERATURE MODULATED DSC (MTDSC) (M. READING)

The use of modulation in calorimetry is very different from its use in thermogravimetry or evolved gas analysis because a change in temperature causes a heat flow due not simply to changes in reaction rate but also from heat capacity. This can be expressed as:

$$dQ/dt = Cp.dT/dt + hf(\alpha)Ae^{-Ea/RT} \qquad (1)$$

where Q = heat

t = time

Cp = heat capacity

h = some constant of proportionality

$f(\alpha)$ = some function of extent of reaction α

$Ae^{-Ea/RT}$ = The Arrhenius equation.

This means that, at any point in time where the heating rate is not zero, the reaction rate cannot be known (and so controlled) unless Cp is known which would generally not be the case. This leaves the option open of using a stepwise method based on successive isothermal plateaus. A method of this type was implemented by Cloudy [40]. The temperature was increased and held isothermal until the heat flow retuned to zero (within the measuring capabilities of the instrument) whereupon the temperature was stepped up by a predetermined amount. This necessarily means that a chemical reaction, for example, must occur within a single step. The next logical extension is to implement the method of Sorenson but before this is discussed the technique of modulated temperature DSC (MTDSC) must be briefly explained.

Modulation has been used in calorimetry since the 1960s in the form of AC calorimetry [41]. In this technique a sample is periodically heated with an IR lamp or electrically and the amplitude and phase of the temperature change is measured. There is one example of this type of experiment with a DSC [42] but it was not until in 1992 that Reading et al. introduced a technique where some underlying temperature program (typically linear or isothermal) was combined

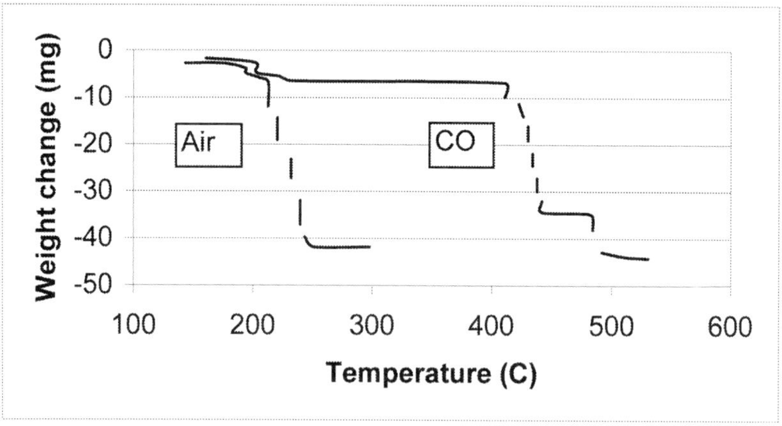

Figure 3-13. FSIA on Ce-carbonate in ait and CO.

with a modulation and the response to the modulation (amplitude and phase) was measured while simultaneously a signal equivalent to conventional DSC was obtained [43–46]. This method then provided the basis for a deconvolution into 'reversing' and 'non-reversing' signals. Figure 3-14 gives the raw data from a MTDSC experiment for PET. Typically deconvolution is achieved by first averaging over the period of a modulation, to produce a signal equivalent to conventional DSC. This signal is subtracted from the raw data before applying a Fourier transform to obtain the phase and amplitude of the heat flow modulation. Essentially this amplitude and the average signal provide two measures of heat capacity that must give the same value when there is no transition. The temperature and rate of change of temperature are given by:

$$T = T_0 + \beta t + B \sin\omega t \qquad (2)$$

and

$$dT/dt = \beta + \omega B \cos\omega t \qquad (3)$$

We can write:

$$\langle dQ/dt \rangle / \beta = C_{p,t} = \text{the total heat capacity} \qquad (4)$$

$$A_{HF} / \omega B = C_{p,r} = \text{the reversing heat capacity} \qquad (5)$$

Where $\langle dQ/dt \rangle$ = the average heat flow over the period of one or more modulations

A_{HF} = amplitude of the heat-flow modulation

A further signal can be calculated:

$$C_{p,t} - C_{p,r} = C_{p,nr} = \text{the non-reversing heat capacity} \qquad (6)$$

If we apply this analysis to the data given in Figure 3-14 we obtain the deconvoluted data shown in Figure 3-15 where the three signals are plotted. Where there is no transition the average heat flow and the amplitude of the modulation in heat flow give the 'true' (completely reversible) heat capacity associated with the molecular motions (principally vibrations and consequently sometimes called the vibrational heat capacity). This means the non-reversing heat capacity must be zero. When a transition occurs the measurement provides apparent heat capacities because there will be some enthalpy associated with the transition such as a heat of reaction. The interesting observation is that the two measures are different when a transition occurs

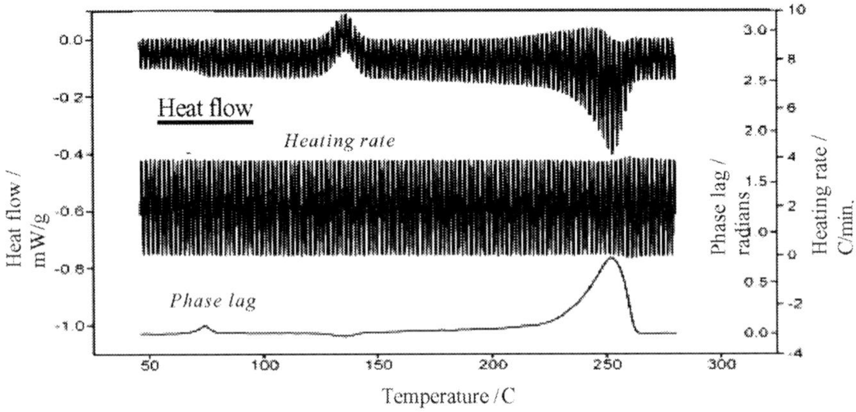

Figure 3-14. Raw data from a modulated temperature DSC experiment on amorphous PET.

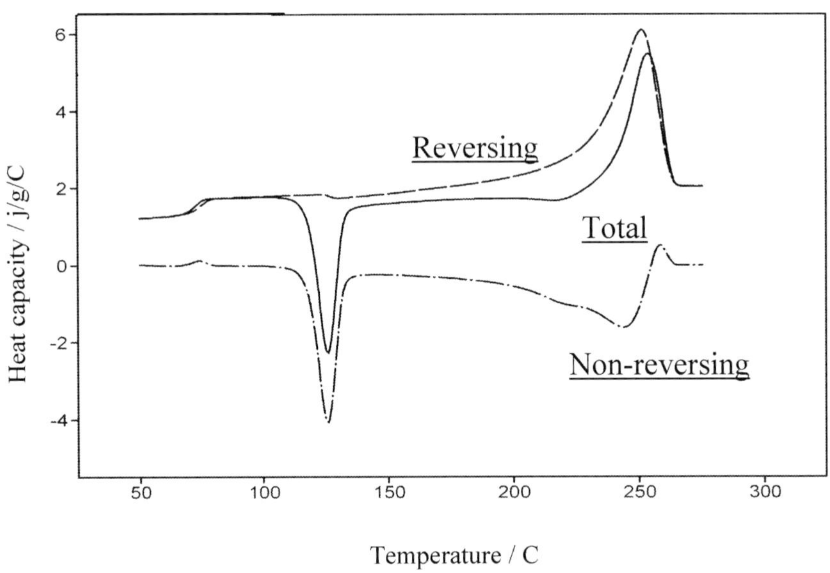

Figure 3-15. The result of the deconvolution of the data shown in Figure 3-14.

The simple's case we can consider is that of some processes, like a chemical reaction, that follows Arrhenius kinetics. We can extend Eq. (8) to obtain:

$$dQ/dt = \beta\, C_p + hf(<\alpha>)Ae^{-E/RT} \qquad \textit{...the average signal}$$
$$+B\omega C_p \cos\omega t + C\sin\omega t \qquad \textit{...the response to the modulation}$$

where $C = Bf(<\alpha>).d(hAe^{-E/R<T>})/dT = Bf(<\alpha>).(hAE/R<T>^2)\, e^{-E/R<T>}$.

For most chemical reactions under most experimental conditions:

$C \cong$

Therefore;

$C_{p,r} = \mathrm{Cp}$

This then means

$$C_{p,nrr} = hf(<\alpha>)Ae^{-E/RT}/\beta \tag{7}$$

In other words it is possible to separate out the heat flow due to heat capacity from that due to the chemical reaction. In Figure 3-15 we see a glass transition, a crystallisation and a melting event. The crystallisation event follows Arrhenius type kinetics. It can be seen that the enthalpy associated with the crystallisation at about 125°C does not appear in the reversing signal and so the latent heat of crystallisation appears in the non-reversing signal in accordance with the analysis presented above. A detailed discussion of the interpretation of the other transitions is beyond the scope of this chapter except to say that, in each case, where there is a transition, the non-reversing signal ceases to be zero. It should be noted that above we have discussed the so-called 'simple' deconvolution procedure that does not use the phase angle. The 'complete' deconvolution procedure, which does use the phase angle, can also be used without substantially changing the arguments [43–46]. It should also be noted that the modulation does not have to be a single sine wave [46,47]. Multiple sine waves, square or triangular waves can be used (a step-iso programme were each cycle is the same is a special case of a triangular wave combined with a linear ramp) which can then always be treated as a multiplicity of sine waves. This then means the reversing signal can be determined for a multiplicity of frequencies simultaneously [46,47]. Alternatively the curve fitting approach of Reading can be used [48].

Price [47] developed a novel technique based on a synthesis of the methods of Sorenson (for the temperature control) and Reading (for the deconvolution) so that the temperature was increased when the heat flow was below as certain level then heating ceased when a lower non-zero threshold was passed. Typically it is not possible in a DSC to rapidly change the temperature and then achieve a perfect isotherm. In reality the temperature is either being increase quickly or slowly. As shown in Figure 3-16, a series of steps result that is equivalent to a modulated heating program based on a triangular modulation where the average heating rate is not controlled and neither is the amplitude or frequency of the temperature modulation.

The question then how these results should be interpreted? Price demonstrated an approach that was based on MTDSC. The details of the deconvolution differ because the stepwise procedure produced 'modulations' with an irregular frequency and amplitude. Price adapted a method proposed by Reading that does not use a Fourier transform but is instead based on a curve fitting technique (details can be found in [48]). The results are shown in Figure 3-17.

When comparing Figures 3-15 and 3-17 the different conventions being used must be taken into account. In Figure 3-15 the results are expressed as heat capacities in Figure 3-17 as heat flows. The relationship is simply that, in the

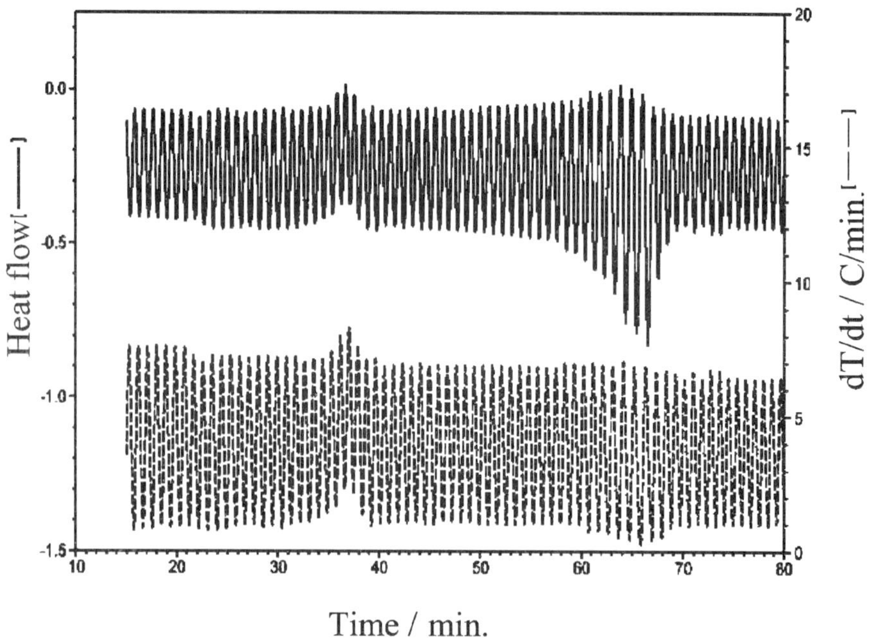

Figure 3-16. Raw data from step-wise DSC for amorphous PET.

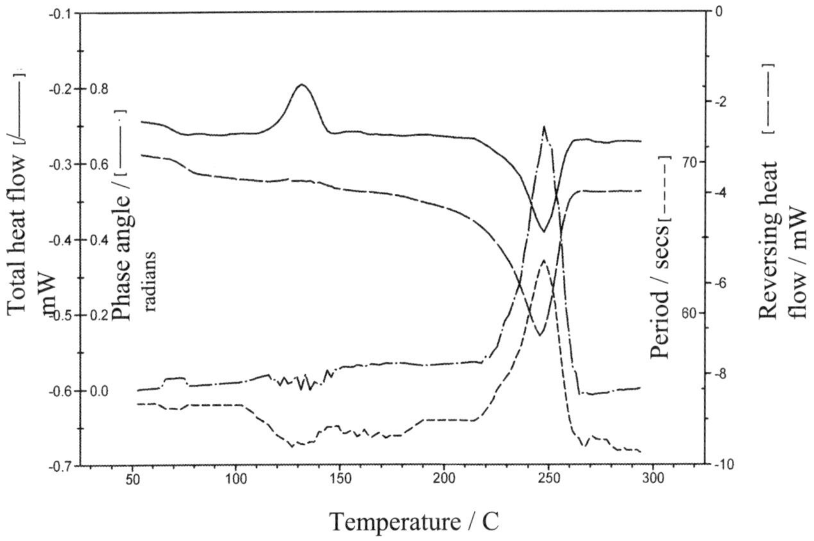

Figure 3-17. Deconvoluted data from Figure 3-16.

case of MTDSC, the heat capacity signals can be converted to heat flows by multiplying by the average (underlying linear) heating rate. Notwithstanding this difference in presentation, the two sets of results are comparable except the modulation period changes significantly during the course of the experiment.

A version of this technique has subsequently become commercially available from Perkin Elmer called step-scan DSC. The there is the option of controlling the threshold values in terms of the rate of change of the heat flow rather than the value of the heat flow but essentially the same result is achieved in terms of a giving rise to a series of temperature steps during the course of a transition which is then equivalent to a modulation with varying amplitude and frequency. In the commercial step-scan the method of deconvolution is also different. Each modulation consists of a temperature increase followed by an isothermal plateau. Integrating the area under the heat flow peak between isothermal plateaus and dividing it by the height of the temperature step gives an approximation to the reversing heat capacity described above provided there are many such steps during a transition. The heat flow during the isothermal plateaus gives the non-reversing heat flow. The disadvantages of this approach to deconvolution are that it provides a single point per modulation (a discrete average) instead of the running average of the Fourier transform or linear fitting deconvolution procedures, it is often not possible to achieve a true isotherm when an energetic transformation is occurring (thus the deconvolution will not be valid) and the time taken to achieve an isotherm always means the steps have long periods relative to what could be achieved with a sine wave.

Using the stepwise method to produce modulations is, therefore, possible but the question is whether it is the best approach. In general it is better that the frequency and amplitude of the modulation remain constant because this makes the interpretation of the data much simpler. An alternative would be to use the non-reversing signal as the trigger for a change in average heating rate. When there is no transition, and so the non-reversing signal is zero, the average heating rate could be fast but when the non-reversing signal deviates from zero, either exothermically or endothermically, the heating rate could be decreased. There are, as yet, no published demonstrations of this technique.

A point that should be noted is that there is a fundamental difference between basing a SCTA method on a reversible equilibrium quantity like heat capacity and a quantity like the enthalpy associated with a, often irreversible, chemical reaction. Heat capacity has, for the purposes of this discussion, no kinetics. By use of modulation it can be measured but to then proceed to 'control' it by, say, maintaining it constant as a function of time, makes no sense, the experiment would simply be a never-ending quasi-isotherm. In contrast it makes perfect sense for a reaction enthalpy. It is true that transformations, including chemical reactions, can cause changes in heat capacity so that it might be possible to envisage an experiment where the rate of change of heat capacity might be controlled. Nevertheless, when looking at SCTA for calorimetry, these distinctions must be born in mind.

3.4.4. THE PROPORTIONAL HEATING TA (PHTA) (PARKES, BARNES, BROWN AND CHARSLEY, 1999)

The great advantage with SIA is that it gives a high resolution, but the price to pay is that these measurements are rather time consuming. In order to overcome this disadvantage Parkes, Barnes, Brown and Charsley [49] proposed the PHTA technique which provides a better resolution in the temperature domain in a given time than linear heating but which also reduces the measuring time considerably. In essence it uses a very low, zero or negative heating rate *during* thermal processes, but saves much time by employing higher rates *between* them. In essence, the PHTA technique smoothly alters the heating rate between pre-set maximum and minimum values in such a way as to provide the optimum balance between resolution in transformation temperature (for which the maximum is sought) and experiment time (for which the minimum is sought).

For a PHTA experiment four pre-set parameters are required: the 'target' reaction rate, a mathematical function relating the reaction rate to the heating rate and maximum and minimum heating rates (the latter may be positive, zero or negative). When no reaction is occurring, PHTA sets the heating rate to its maximum value. However, if the reaction rate equals or exceeds the pre-set target, the heating rate is reduced to its minimum value. In between these two boundaries the heating rate is continuously varied as a function of the reaction rate, the simplest routine for which is a linear relationship, but many other functions are possible. However, procedures that are not smooth, i.e. those with discrete steps, will result in heating rate changes which are themselves discontinuous, which is undesirable.

The operation of this method can be seen in Figure 3-18 which shows a dehydration of copper sulphate pentahydrate.

The experiment starts with the maximum heating rate, set here at 5°C/min. At a temperature of ca. 50°C the reaction commences and, as it increases in rate, the heating rate decreases. When the reaction rate reaches, and then exceeds, the target level (which in this example is 1000 hygrometer arbitrary units), the

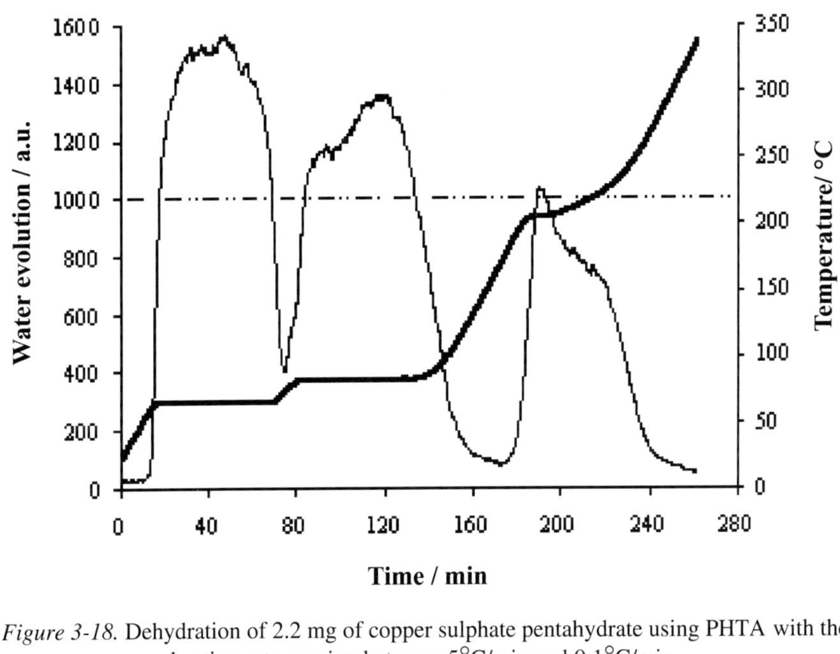

Figure 3-18. Dehydration of 2.2 mg of copper sulphate pentahydrate using PHTA with the heating rate varying between 5°C/min and 0.1°C/min.

heating rate is at its minimum value of 0.1°C/min. Finally, the first step of the process approaches completion and its rate declines, leading to a corresponding increase in the heating rate to 5°C/min again, until the second dehydration stage begins and a similar sequence of events occurs. The third dehydration stage only transiently acquires a reaction rate sufficient to reach the target level and hence the heating rate is faster through it than for the other two stages. This demonstrates the limitation of having to use a pre-set target reaction rate which may not be suitable for all the events in a process involving several stages. Moreover, the shape of the EGA profile in PHTA is dependent on both the reaction kinetics and the combination of heating rates with the function relating reaction rate to heating rate.

In addition to providing enhanced resolution, PHTA has the additional advantage of being a versatile technique, as alterations in the control parameters can be used to change the nature of the experiment. If the minimum heating rate in the experiment is set to zero and a 'step' function relating the reaction and heating rates is used, the PHTA technique becomes in effect SIA. However although an improvement in the resolution is clearly obtained by using this form of the PHTA technique, better resolution can be obtained using dedicated SIA, as shown in Figure 3-11.

3.4.5. THE "MAX RES" TECHNIQUE (SCHENKER AND RIESEN, 1997 [50]) (M. READING)

This method also results in a progressive decrease or increase in heating rate in response to changes in the rate of weight loss. The algorithm that is used is illustrated schematically in Figure 3-19.

The operator chooses the starting heating rate, a maximum and minimum heating rate and a factor together with the upper (maxrate in Fig. 3-19) and lower (min rate in Fig. 3-19) threshold values for the rate of weight loss. The operator also chooses a time-out period that puts a minimum time that will elapse before any change in behaviour by the sample will result in a change in heating rate. The experiment commences with the starting heating rate, typically the rate of weight loss will be zero. After the time out period the programmed heating rate is increased by multiplying it by the factor. If the rate of weight loss is still below the upper threshold then after the second time out period the heating rate is again increased by multiplying it by the factor and so on until either the rate of weight loss exceeds the upper threshold or the increment in heating rate would take it above the set maximum in which case it remains unchanged. When the maximum rate of weight loss is exceeded then the same process is employed except now the heating rate is divided by the factor after each time out period. Results for this method are compared with a conventional

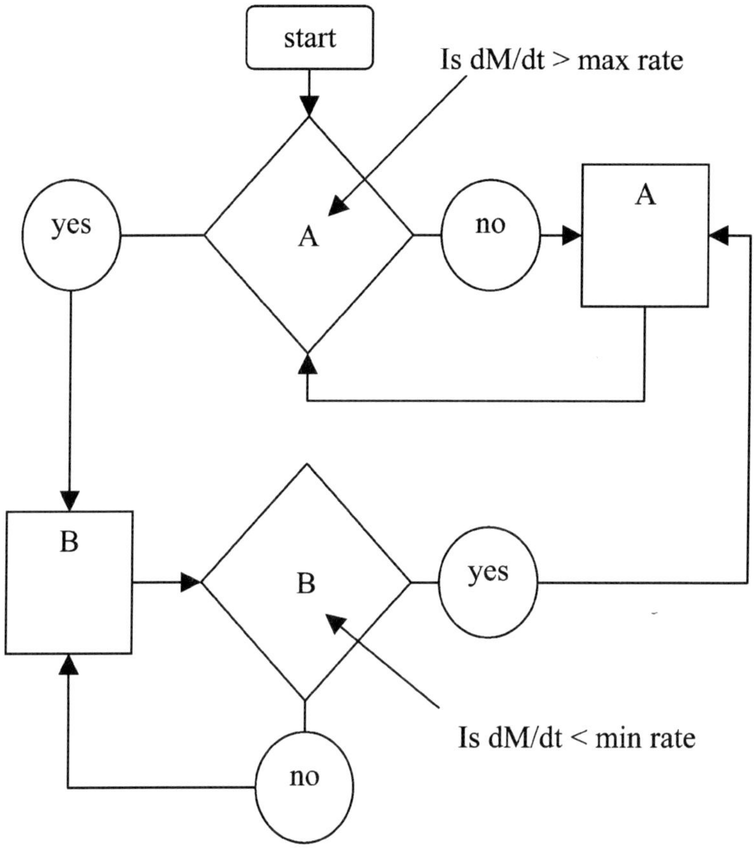

Figure 3-19. Algorithm used in the maxres method. A = multiply the heating rate by an operator chosen factor (unless this exceeds the operator selected maximum heating rate in which case it is held constant) then wait for the time-out period. B = divide the heating rate by the factor (unless this is then smaller than the operator selected minimum heating rate in which case it is held constant) then wait for the time-out.

linear rising temperature experiment in Figure 3-20. In a similar way to the dynamic rate method, the maxres approach shows an apparent increase in resolution.

Again the reaction rate is not controlled in the sense of following a predetermined function of time. For both the dynamic rate and maxres methods neither rate nor temperature follow a predetermined function of time. It is interesting to note that both cases it is difficult to see how these approaches could have been implemented before the advent of computer control.

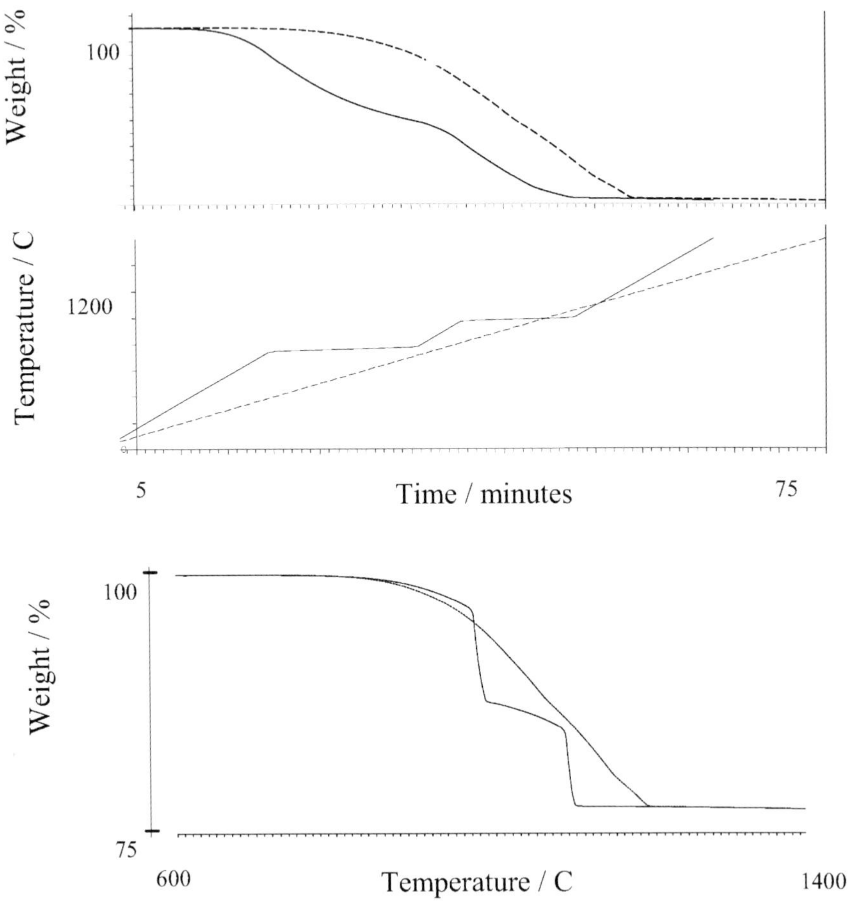

Figure 3-20. Comparison of linear rising temperature (dashed line) and Maxres results (solid line) for a mixture of $BaCO_3$ and $SrCO_3$.

3.4.6. THE DYNAMIC HEATING TECHNIQUE (OPFERMANN) (M. READING)

Dynamic heating proposed by Opfermann [54] where the algorithm regulates the heating rate during the reaction in such a way that the rate is high in the beginning and at the end of the reaction but slow during the middle part of the reaction where the rate of transformation is high. This technique, which is designated by Netzsch as Dynamic TA, both shorten the overall measuring time and improves the resolution.

3.5. Combined Rate and Temperature-Controlled Thermal Analysis

In contrast to the techniques discussed in the previous sections, a characteristic feature of the combined techniques is that neither the temperature nor the reaction rate follow a predetermined function. Or expressed more precisely, that the algorithm determines a smooth and continuous response to the sample behavior but does not provide for either temperature or reaction rate to be a predetermined function of time. For this technique Reading [55] proposed the term: Constrained Rate Thermal Analysis based on the fact that the sample is allowed only to vary within certain overall preset limits. This, however, is not only the case in SCTA but also in some other TA techniques and in this book we therefore prefer the term Combined SCTA.

In this section we shall focus on the Dynamic Heating Rate Thermal technigue introduced by Sauerbrunn and Gill.

3.5.1. HIGH RESOLUTION THERMOGRAVIMETRY (GILL, SAUERBRUNN AND CROWE, 1992) (M. READING)

B.S. Crowe and colleagues proposed an method in which the heating rate is related to the rate of weight loss by the equation [56]:

$$dT/dt = \beta_{max} / \exp(((d\alpha/dt\ 4^{R}) / 256)^{S}) \tag{8}$$

where β_{max} = selected maximum heating rate
α = fractional weight loss, expressed as a percentage
R = resolution, usually between 0 and 8
S = sensitivity, usually between 1 and 8

The heating rate is never allowed to fall beneath 0.001C/min.

The logic behind this approach is simply that, when nothing is happening the heating rate can be rapid so that time is not wasted and when weight loss begins to occur, the heating rate is slowed so that resolution is improved [36]. Examples of the curves this function generates can be seen in Figure 3-21 and how this influences the progress of temperature and rate of weight loss as a function of time is given in Figure 3-22.

Figures 3-23 and 3-24 give examples of how the variables S and R influence the outcome of experiments and compares the results with the conventional linear rising temperature method. In all cases the dynamic rate method shows the two steps in this reaction as more clearly delineated. This is simply because the heating rate is slow during the major weight losses and so they occur over a narrower temperature interval. As the rate declines, the temperature is increased rapidly thus forcing, in the temperature domain, a clear step between the first and second weight losses. This 'increase' in resolution when compared to a linear temperature ramp experiment can be achieved even when the dynamic rate experiment that take significantly less total time.

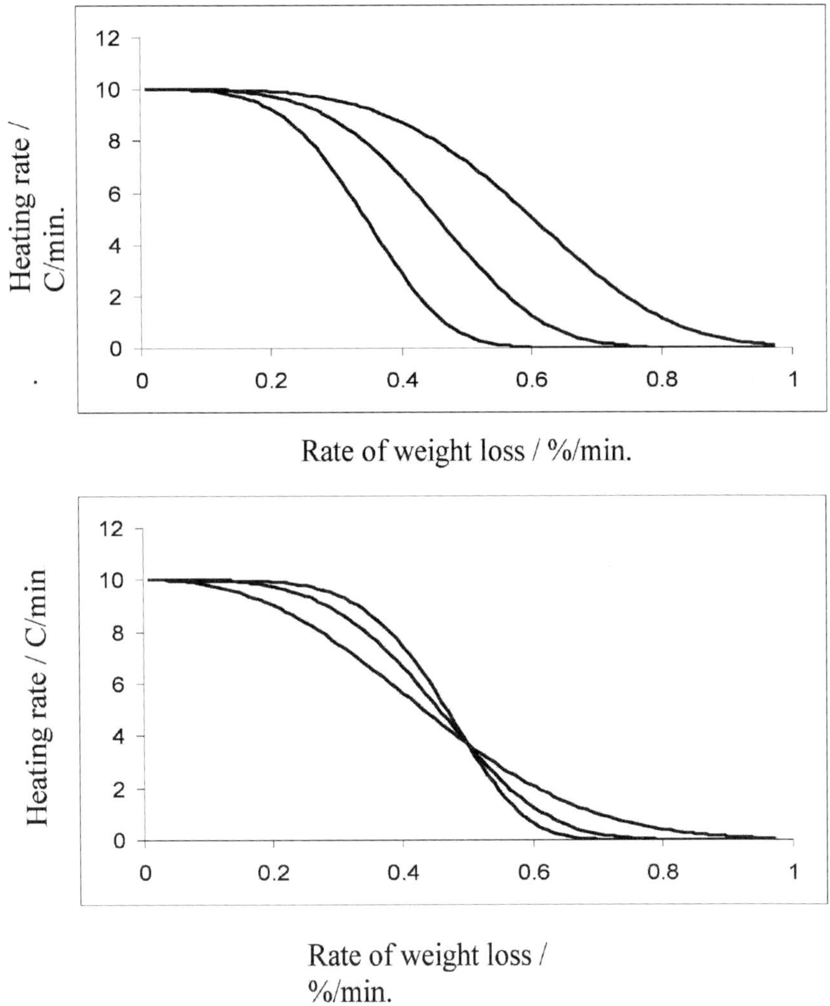

Figure 3-21. Graphs illustrating the relationship between heating rate and rate of weight loss in the 'dynamic rate' technique. The relationship is give by equation 1 using parameters, top R = 4.3, 4.5 and 4.7 S = 4, below R = 4.5 and S = 2.5, 4 and 5.5.

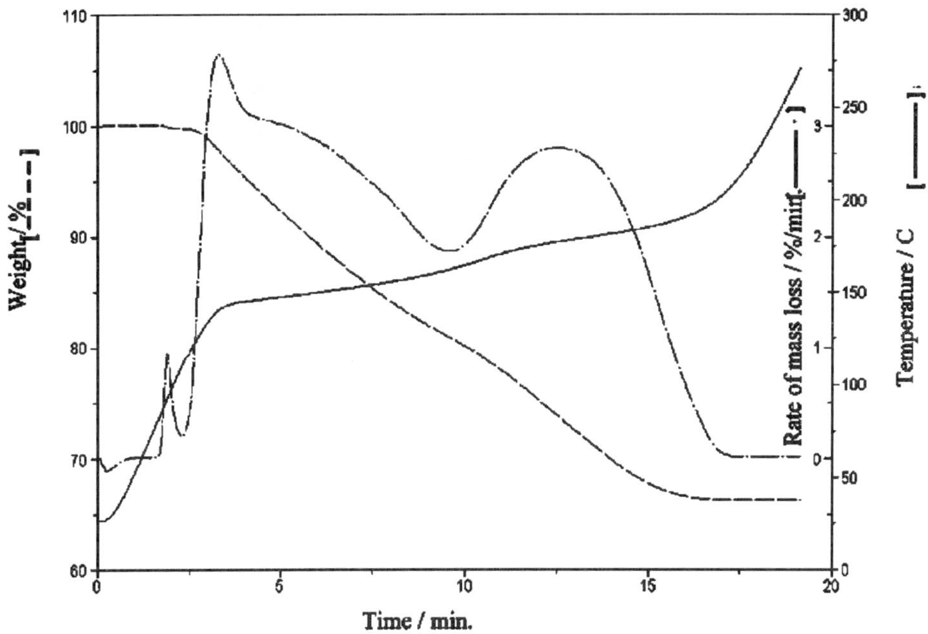

Figure 3-22. Results for a dynamic rate experiment showing how temperature, weight and rate of weight loss changes with time for a 50/50 mixture of $KHCO_3$ and $NaHCO_3$.

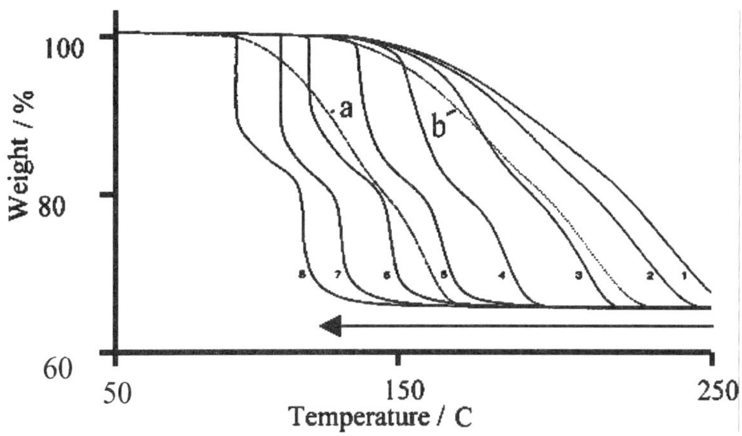

Figure 3-23. Illustration of the effects of changing the resolution setting for a 50/50 mixture of $KHCO_3$ and $NaHCO_3$. The dashed curves a) and b) are conventional linearly rising temperature experiments at 1 and 20 C/min. respectively. The dynamic rate settings were 50 C/min., a sensitivity of 1 and resolution settings of 1 to 8 in the direction indicated by the arrow.

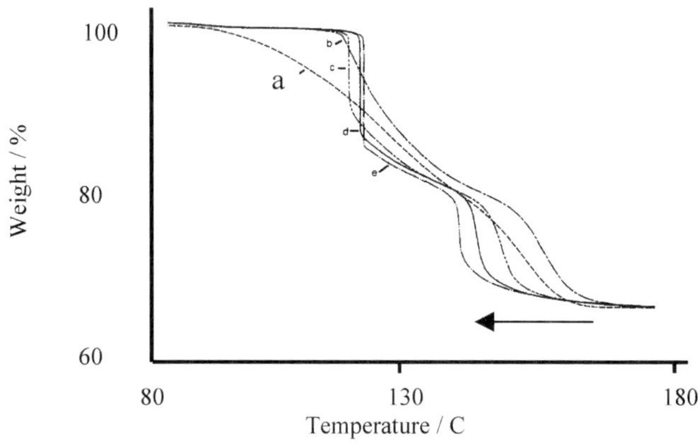

Figure 3-24. Illustration of the effects of changing the sensitivity setting for a 50/50 mixture of $KHCO_3$ and $NaHCO_3$.The curve labelled a) is a conventional linearly rising temperature experiment at 1 C/min. The dynamic rate settings were 20 C/min, a resolution of 5.0 and sensitivity settings of 1, 2, 4 and 8 in the order indicated by the arrow.

When might this method be preferred over constant rate thermal analysis? One answer to this is that one method might consistently provide superior resolution compared to the other. In fact it is not clear that this is the case. Some way of defining resolution must be decided upon before quantitative comparisons can be made. This question of assessing resolution will be addressed toward the end of this chapter. One reason for favouring the constant rate approach is that, assuming that a purge gas at a constant flow rate sweeps the sample chamber, it maintains the partial pressure above the sample bed constant. This is possibly a significant advantage when studying kinetics as this parameter might reasonably be expected to influence the reaction rate and so needs to be as well defined and controlled as possible. When maintaining the rate of weight loss constant, good control is easier to achieve at relatively slow reaction rates. This makes the experiments very long, which is a disadvantage, particularly in industrial laboratories. Increasing the rate degrades the quality of the control and so the ability to discern subtle transitions. The dynamic rate method offers a good alternative as it achieves apparent improvements in resolution with rapid experiments.

3.6. Sample Controlled Thermomicroscopy (E.L. Charsley, C. Stewart)

Sample controlled thermal analysis techniques such as constant rate transformation analysis or stepwise isothermal analysis, where the transformation rate of the sample itself is used to control the experiment, are becoming increasingly important [57]. The measurements are normally carried out using changes in the sample mass, sample dimensions or in the evolved gas, as the property used to control the experiment, and enable reactions to be studied in greater detail than is possible using linear heating techniques. A new approach is described here where a thermomicroscopy system has been developed to enable the intensity of the light reflected or transmitted by the sample to be used as the controlling signal [58].

3.6.1. EXPERIMENTAL

A block diagram of the system is shown in Figure 3-25. The measurements have been carried out using a Linkam Model THMS600 hot stage. This unit operates over the range –180°C to 600°C and uses a silver block heater in conjunction with a liquid nitrogen cooling system. Samples can be run in a controlled atmosphere and the fast response of the system makes it ideal for sample controlled studies. The intensity of the light signal from the sample was monitored by a silicon photodetector fitted into one eyepiece of a Nikon Labophot microscope. The standard temperature control system has been modified so that the sample temperature is controlled by the rate of change of the light intensity signal. The light intensity and sample temperature signals were recorded using a computer fitted with a Strawberry Tree 16-bit dynamic range data acquisition board.

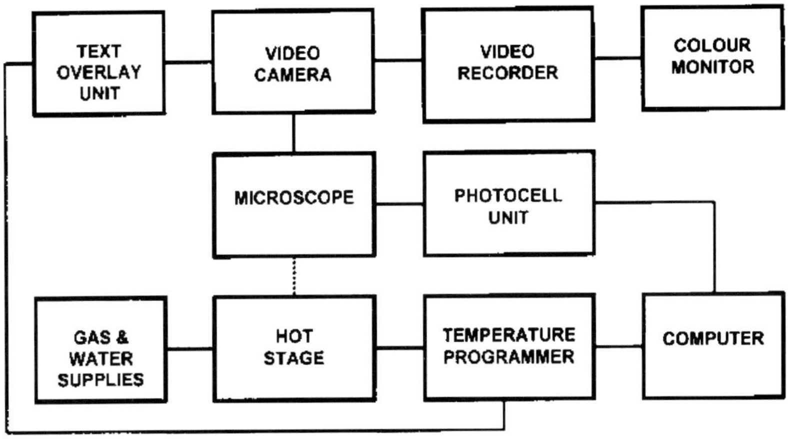

Figure 3-25. Block diagram of sample controlled thermomicroscopy system.

Reflected light intensity measurements (RLI) were performed using white light [59]. In addition, depolarised light intensity (DLI) measurements have been carried out with transmitted polarised light. Crossed-polars were used so that the only light transmitted was due to the rotation of the polarised light by the crystalline structure of the sample [60,61]. A number of control strategies have been implemented including constant transformation rate measurements and stepwise isothermal analysis.

3.6.2. RESULTS AND DISCUSSION

Decomposition of Silver Oxide Under Reflected Light Conditions

The decomposition of silver (I) oxide was chosen since it produces a well-defined colour change from black to white on decomposition to form the metal. An experiment carried out under constant rate (CR) conditions is shown in Figure 3-26.

The sample temperature can be seen to reduce rapidly once the reaction has started. This is typical of a nucleation reaction and was also observed in sample controlled studies where oxygen evolution was used to control the rate of decomposition. The sample temperature is then maintained essentially constant to enable a linear rate of reaction to be established over the complete decomposition reaction. Figure 3-27 shows a comparison of the sample controlled experiment with an experiment made under linear heating conditions at 10°C min–1 and illustrates the markedly lower temperature of reaction in the former case.

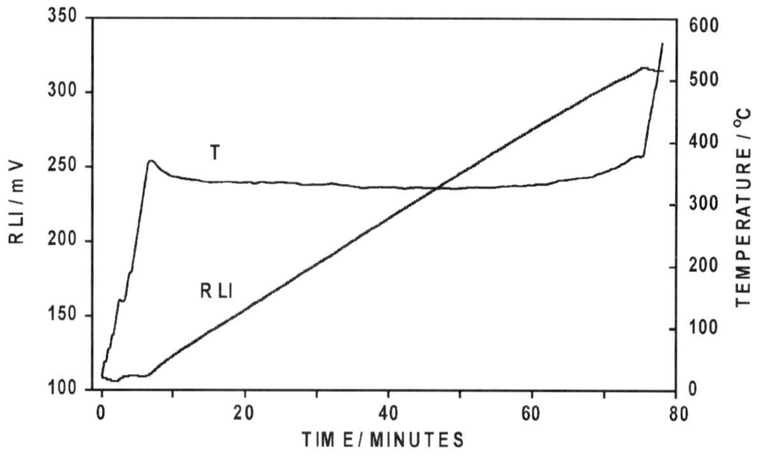

Figure 3-26. Constant rate reflected light intensity curves for the decomposition of silver (I) oxide.

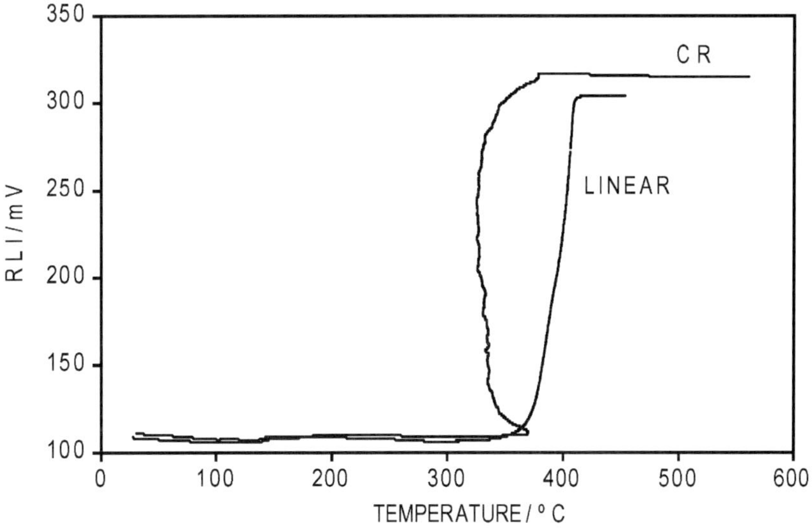

Figure 3-27. Constant rate & linear heating reflected light intensity curves for the decomposition of silver (I) oxide.

Transmitted Light Intensity Measurements

The ability to study phase transformation under polarised light offers the potential to apply sample controlled techniques to a range of tranformations which do not take place with a change in mass. These include fusion, solid-solid transitions and recrystallisation reactions.

This is illustrated in Figure 3-28 by a study of the behaviour of a crude oil on cooling under constant rate conditions. As the sample is cooled below room temperature the DLI signal begins to increase due to wax formation. In order to maintain a constant rate of wax formation the sample has to be cooled at a linear rate. This suggests that the amount of wax formed is directly related to the temperature and that there are no large hysteresis effects. Similar results were obtained using a DLI-stepwise isothermal approach.

In conclusion, a new range of sample controlled thermomicroscopy techniques have been developed which enables samples to be studied under reflected light or transmitted light conditions. The ability to make depolarised light intensity measurements should be of particular benefit in studying transitions in a wide range of materials including polymers, pharmaceuticals and liquid crystals.

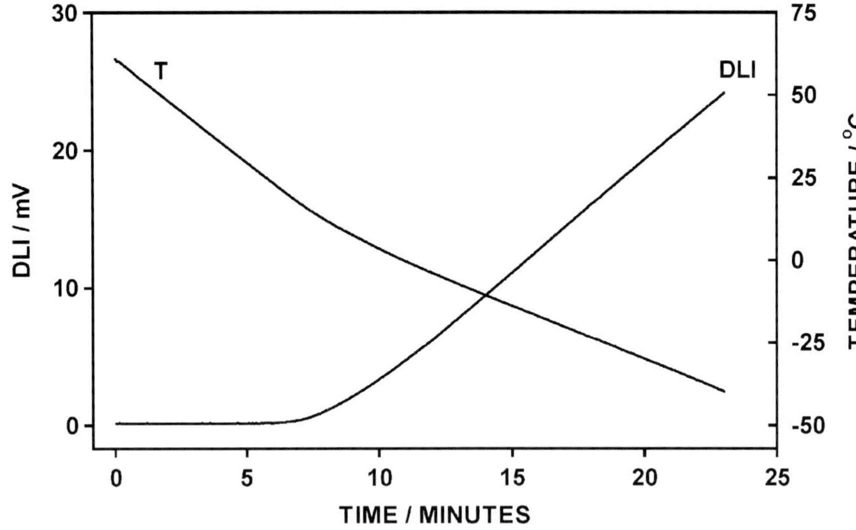

Figure 3-28. Constant rate depolarised light intensity curves obtained under cooling conditions for a crude oil.

3.7. Sample-Controlled Reaction Rate by Gas Blending Techniques (P.A. Barnes, G.M.B. Parkes)

As pointed out in Chapter 1, SCTA embraces the branch of thermal analysis where a feed-back from the sample is used to control its heating or cooling. This possibility of using a feed-back from the sample to control its rate of transformation was extended by Parkes, Barnes and Charsley to a type of experiment which is out of SCTA proper but which can be considered as closely related to it, both because of its origin and because of its application [62]. In this approach, which is basically isothermal, the reaction rate, which is programmed in advance (as a function of time), is achieved through the appropriate control of the concentration of the reactive gas-phase component(s). This is not a SCTA experiment proper, (i) because it is not a thermal analysis experiment (since there is no recording of any physical parameter *vs.* temperature) and (ii) because the feed-back from the sample does not act upon the heating or cooling of the sample. Now, like a SCTA experiment, it does use a feed-back from the sample to control a reaction rate (for instance, to keep it constant) and its applications and interpretation have much to do with SCTA, as explained hereafter. Also, as will be seen at the end of this section, this technique can be directly associated, in some cases, with a real SCTA experiment.

The technique combines two or more mass-flow controllers in parallel to supply the gases to the sample. Via on-line computer control, they govern the relative proportions of the reactant gas(es) and the carrier gas. The combination is configured to allow the production of any blend of the gases, at a total constant flow rate, in whatever way is required to force the reaction rate to obey the requirements of the SCTA method selected. For example, gas blending analogues of conventional linear heating, constant reaction rate, PSH and PHTA can be implemented. The gas blending approach is particularly suited to emulating process such as TPR and TPO [63]. Though the technique is, basically, isothermal, it should be noted that series of isothermal steps can be employed if required.

A typical system for this approach and also for SCTA, which illustrates the alternative feedback loops, is shown in Figure 3-29.

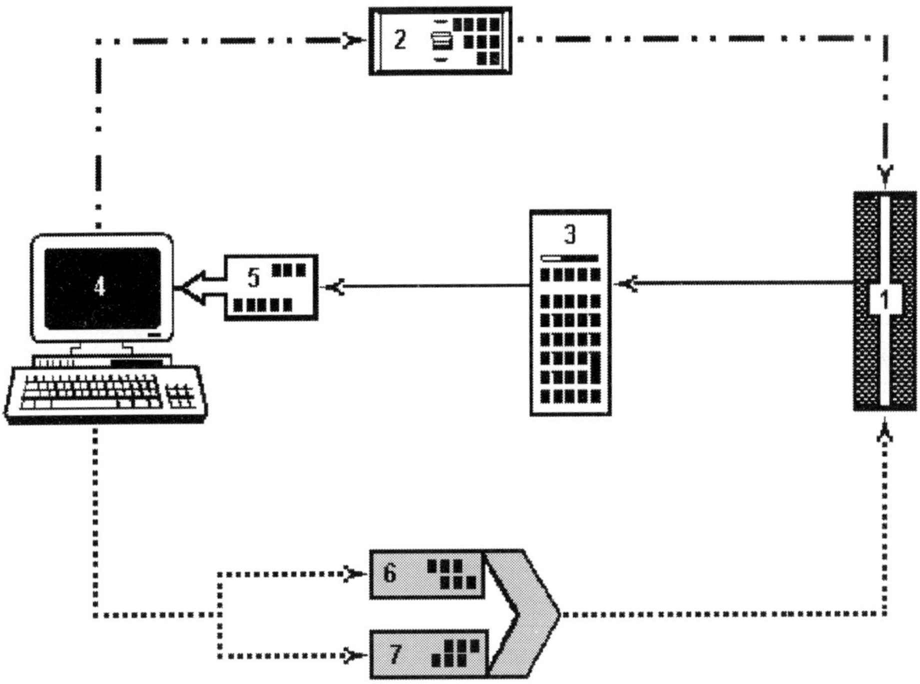

Figure 3-29. Schematic diagram of the SCTA technique showing two alternative feedback loops (temperature and concentration): (1) furnace, (2) Eurotherm, (3) detector, (4) computer, (5) ADC, (6) and (7) mass flow controllers.

In both cases, the monitored level of a selected product gas is used as a measure of the rate of a reaction and controls either the sample temperature or gas composition, respectively, in such a way as to give the desired reaction profile. The principle components include a furnace (1), the temperature of which is continuously altered (in case of a SCTA experiment) via a temperature controller (Eurotherm, 818P) (2), which operates via software specifically developed to implement a range of SCTA regimes [64]. The reaction rate is monitored by measuring the evolved gases using a suitable detector (3), the output of which is fed back to the computer (4) via a 16-bit ADC (5). In SCTA techniques, such as CRTA, a single mass-flow controller (6) is used to set the flow of gas(es) through the furnace. For gas blending experiments, a second mass-flow controller (7) is added. The software now operates the mass-flow controllers to produce the required (continuously varying) blend of carrier and reactant gases whilst keeping the total flow rate constant. The feedback loop incorporates a 3-term PID controlled algorithm in the software to provide optimum control of the reaction rate for both SCTA and the Gas Blending technique.

Theory

The theory of the Sample Controlled Gas Blending approach is analogous to that of SCTA and is based on a general expression for the rate of a solid-gas reaction:

$$d\alpha / dt = Ae^{-E_a / RT} f(\alpha) \tag{9}$$

where the variables A and E_a can be regarded as temperature coefficients, and the function of concentration, which is dependent on the reaction, can be represented generally as:

$$f(\alpha) \approx c^n \tag{10}$$

where the variable *n* usually assumes values in the range 0–2.

It can be seen that, in the case of SCTA experiments, the overall shape of the thermo-analytical curve is determined largely by the exponential term in equation (1). However, if the experiment is carried out isothermally, the curve shape will be governed by the concentration term defined in equation (2). The important consequence of this is that, in general, a change in temperature exerts more influence than variations in concentration of a gas-phase reactant.

The effect of this is that peaks found in conventional thermal analysis (i.e. those produced under conditions of a linear increase in temperature) are generally sharper than those found in the analogous gas blending experiment, where the concentration of a gas-phase reactant increases linearly under isothermal conditions. However, if the concentration is altered such that it produces a constant reaction rate, the reaction proceeds under near equilibrium conditions of both temperature *and* concentration.

It is clear that the rate of a suitable reaction can be controlled, in principle, by both temperature and concentration parameters. However in such a case, interpretation of the resulting profiles becomes complicated as the reaction rate is influenced by the combined temperature and concentration contributions. A more practical approach in the study of a series of consecutive reactions is to link the temperature and gas-phase concentration control elements to provide a series of increasing isothermal plateaus. In this case, one enters the scope of Stepwise Thermal Analysis, i.e. the scope of SCTA proper. Each of the processes can then be the subject of a Sample-Controlled Gas Blending experiment, whilst operating at its own isothermal temperature, so optimising the conditions for each reaction.

References

1. C.S. Smith, Trans. A.I.M.E. (Metal Division) 137 (1940) 23.
2. B. Boinon, Thesis, Université Claude Bernard, Lyon, 10th June 1974, No. d'ordre 257.
3. M. Reading, in "Thermal Analysis, Techniques and Applications" E.L. Charsley and S.B. Warrington (Eds.), Royal Society of Chemistry, Cambridge (1992) 126.
4. J. Rouquerol, Bull. Soc. Chim. Fr. (1964) 31.
5. J. Rouquerol, Thesis, Faculty of Sciences of Paris University, 19 November 1964 (Série A, No. 4348, No. d'ordre 5199).
6. F. Rouquerol, J. Rouquerol and K.S.W. Sing, "Adsorption by powders and porous solids: principles, methodology and applications", Academic Press, London, New-York (1999) 467 p.
7. J. Rouquerol, Thermochimica Acta 300 (1997) 247.
8. J. Rouquerol, J. Thermal Analysis 2 (1970) 123.
9. J. Mayet, J. Rouquerol, J. Fraissard and B. Imelik, Bull. Soc. Chim. France 9 (1966) 2805.
10. P. Duran, J.L.G. Fierro, and M. Rodriguez, Meas. Sci. Technol. (1990) 523.
11. J. Rouquerol, J. Thermal Analysis, 5 (1973) 203.
12. M. Ganteaume and J. Rouquerol, J. Therm. Anal. 3 (1971) 413.
13. J. Rouquerol, in "Thermal Analysis", D. Schwenker (Ed.), Academic Press, Vol. 1 (1969) 281.

14. M.H. Stacey in "Proceedings 2nd European Symposium on Thermal Analysis" (Aberdeen), D. Dollimore (Ed.), Heyden, London (1981) 408.
15. M.H. Stacey, Anal. Proc. 22 (1985) 242.
16. M. Reading and J. Rouquerol, Thermochimica Acta 85 (1985) 299.
17. J. Arellano, J.M. Criado and C. Real, Thermochimica Acta 134 (1988) 365.
18. C. Real, M.D. Alcala and J.M. Criado, J. Thermal Analysis 38 (1992) 797.
19. P.A. Barnes, G.M.B. Parkes, D.R. Brown and E.L. Charsley, Thermochimica Acta, 269/270 (1995) 665.
20. J. Rouquerol and J.M. Fulconis, ICTAC 11 Book of Abstracts, P. Gallagher (Ed.), Philadelphia (1996) 272.
21. J. Paulik and F. Paulik, Anal. Chim. Acta 56 (1971) 328.
22. L. Erdey, F. Paulik and J. Paulik, Hungarian Patent No. 152197, registered 31 October 1962, published 1 December 1965.
23. J. Paulik and F. Paulik in "Comprehensive Analytical Chemistry", G. Svehla (Ed.), Vol. XII, "Thermal Analysis", Part A, "Simultaneous Thermo-analytical Examinations by Means of the Derivatograph", Elsevier, Amsterdam (1981) 277 p.
24. F. Paulik, "Special Trends in Thermal Analysis", John Wiley and Sons, Chichester (1995) 459 p.
25. H. Palmour and D.R. Johnson, in "Sintering and Related Phenomena", G.C. Kuczynski, N.A. Hooton and C.F. Gibbon (Eds.), Gordon and Breach, New-York (1967) 779.
26. J. Paulik and F. Paulik, Proc. 5th Int. Conf. Thermal Analysis (Kyoto), Heyden, London, 1 (1977) 75.
27. F. Rouquerol and J. Rouquerol in "Thermal Analysis", H.G. Widemann, (Ed.), Vol. 1, Birkhäuser, Basel Vol. 1, (1972) 373.
28. S. Bordère, F. Rouquerol, J. Rouquerol, J. Estienne and A. Floreancig, J. Thermal Analysis, 36 (1990) 1651.
29. J. Rouquerol, Pure and Applied Chemistry 57 (1985) 69.
30. G. Thevand, F. Rouquerol, and J. Rouquerol in "Thermal Analysis", B. Miller, (Ed.), John Wiley and Sons, New-York Vol. 2 (1982) 1524.
31. M.J. Tiernan, P.A. Barnes an G.M.B. Parkes, J. Phys. Chemistry B, Vol. 103 No. 33 (1999) 6944.
32. J. Rouquerol, S. Bordère and F. Rouquerol, Thermochimica Acta, 203 (1992) 193.
33. A. Ortega, L.A. Pérez-Maquéda and J.M. Criado, J. Thermal Anal. 42 (1994) 551.
34. R.L. Blaine, Proceedings of NATAS Meeting, R.J. Morgan (Ed.) (1997) 485.
35. R.L. Blaine, American Laboratory, January 1998, 21.
36. S.R. Sauerbrunn, B.S. Crowe and M. Reading, Proc. 21st North American Thermal Analysis Conference (1992) 137.
37. O.T. Sorensen, J. Thermal Anal. 13 (1978) 429.
38. G.M.B. Parkes, P.A. Barnes, E.L. Charsley, Anal. Chem., 71 (1999) 2482–2487.
39. O.T. Sorensen, J. Thermal Anal. 38 (1992) 213.

40. J.C. Cloudy and J.M. Commerceron, Letoffe, *Thermochim. Acta*, 128 (1988) 251.
41. Kraftmakher, *Zh. Prikl. Mekh. & Tekh. Fiz.*, 5 (1962) 176.
42. Goldbrecht, K. Hamann and G. Willers, *Journal of Physics E: Scientific Instruments*, 4 (1971) 21.
43. Reading, D. Elliott and V.L. Hill, *Proc. NATAS* (1992) 145.
44. Reading, *Trends in Polymer Science*, 1 (1993) 8.
45. A.A. Lacey, C. Nikolopoulos, and M. Reading, 50 (1997) 279.
46. P. Haines, C. Keattch and M. Reading, Differential Scanning Calorimetry, *Handbook of Thermal Analysis*, (Ed.) M. Brown, Elsevier (1998).
47. D.M. Price, *Thermochim. Acta*, 315 (1998) 11.
48. M. Reading, US Patent 5,474,385 (1995).
49. Reading, US Patent 5,474,385 (1995).
50. G.M.B. Parkes, P.A. Barnes and E.L. Charsley, *Analytical Chemistry*, 71 (13) (1999) 2482–2487.
51. B. Schenker and R. Riesen. User Com. Information for Users of Mettler Toledo Thermal Analysis Systems, Dec. 1997, 10–12.
52. G.M. Parkes, et al. Thermochimica Acta, 354 (2000) 39–43.
53. M. Reading, *Handbook of Thermal Analysis and Calorimetry (Chapter 8)*, (Ed.) M.E. Brown, 1998: Elsevier. G.M.B. Parkes.
54. G.M.B. Parkes, P.A. Barnes, and E.L. Charsley, Analytical Chemistry. 71 (13) (1999) 2482–2487.
55. J. Opfermann, Netzsch Geraetebau, Germany-private communication.
56. M. Reading, Thermal Analysis – Techniques and Applications, (Eds.) E.L. Charsley and S.B. Warrington, Royal Society of Chemistry, Cambridge.
57. P.S. Gill, S.R. Sauerbrunn and B.S. Crove, J. Thermal Anal., 38 (1992) 255–266.
58. P.A. Barnes, G.M.B. Parkes and E.L. Charsley, Anal. Chem., 66 (1994) 2226.
59. E.L. Charsley, C. Stewart, P.A. Barnes and G.M.B. Parkes, Paper presented at the 11th International Congress on Thermal Analysis and Calorimetry, Philadelphia, USA, 1996.
60. J.H. McGill Polymer, 2 (1961) 221.
61. E.L. Charsley and A.C.F. Kamp. Thermal Analysis, Vol. 1, (Ed.) H.G. Wiedemann, Birkhauser Verlag, 1972.
62. E.L. Charsley, A.C.F. Kamp and J.A. Rumsey, Thermal Analysis, Vol. 1, (Ed.) H.G. Wiedemann, Birkhauser Verlag, 1980, 285.
63. G.M. Parkes, P.A. Barnes, and E.L. Charsley, Thermochimica Acta, 1998, 320 (1–2) p. 297–301.
64. E.A. Dawson et al., Thermochimica Acta, 335 (1–2), (1999) 141–146.

Chapter 4

SCTA AND KINETICS

J. M. CRIADO and L. A. PÉREZ-MAQUEDA

Instituto de Ciencia de Materiales de Sevilla, Centro Coordinado C.S.I.C. – Universidad de Sevilla; Avda. Américo Vespucio s/n, 41092 Sevilla, Spain

4.1. Introduction

It has been shown in other chapters of this book that different SCTA methods have proved to be very useful in kinetic analysis of solid state reactions and sintering of ceramic materials. The scope of this chapter is to study the fundamental principles of the kinetic analysis of solid state reactions in connection with their application to the kinetic analysis of SCTA curves. This analysis will be mainly focused to the Constant Rate Thermal Analysis (CRTA) method which has been the most widely used for kinetic studies of solid state reactions.

4.2. The Fundamental Problem with the Kinetics of Heterogeneous Reactions

The fundamental problem with heterogeneous kinetics stems from the double process leading to the transformation of the solid, namely the nucleation and the growth of nuclei.

The nucleation, which is the formation of nuclei of the product, results in the *formation* of a reaction interface, whereas the growth of nuclei results in the *displacement* of the previous reaction interface. In the general case, the advancement of the reaction directly results from the complex superimposition of the two above processes, since the growth of nuclei starts as soon as the first nuclei are formed and therefore competes with the formation of new nuclei in the non-reacted parts of the sample. Due to this complexity, parameters T (temperature), p (pressure) and α (degree of reaction) are not independent from each other.

In addition, in conventional, temperature-controlled experiments, the rates of mass and heat transfer are, by principle, uncontrolled, so that, finally, the macroscopic rate of transformation is the consequence of the latter uncontrolled physical parameters and of the former, overlapping, chemical processes. It follows that each experiment is a special case which depends on the reactor (shape and nature of the crucible or pan, thermal coupling with the heating device, presence and type of the cover) and on the other experimental conditions (vacuum or other pressure, carrier or reacting gas, flow-trough system, heating rate or process adopted to start an isothermal reaction ...).

In this context, the major interest of SCTA comes from the possibility to control and lower the rate of advancement of the reaction down to such a value as to make negligible the temperature and pressure gradients within the sample. The experiment then becomes more meaningful since one is now entitled to speak of *the* sample temperature and of *the* surrounding pressure and gas composition. Furthermore, these well-defined experimental conditions can be selected in such a way as to largely favour either the nucleation or the growth of nuclei. In these conditions, one gets a limiting case for which the rate law can be expressed simply, since parameters T, p and α are now independent from each other.

It is precisely a number of limiting cases, for which the Arrhenius law holds and which can be achieved more easily with the help of SCTA which are examined in the following.

4.3. The Fundamental Problem with Non-Isothermal Kinetics: The Expression of the Reaction Rate

It is well known that, in the limiting cases where the parameters T, p and α can be separated and the forward reaction is far from equilibrium, the fundamental equation governing the kinetics of solid state reactions under isothermal conditions is

$$\frac{d\alpha}{dt} = kf(\alpha) \qquad (1)$$

that can be integrated in the form

$$g(\alpha) = \int_0^\alpha \frac{d\alpha}{f(\alpha)} = kt \tag{2}$$

where α is the reacted fraction at the time *t*, f(α) and g(α) are functions depending on the kinetic model as shown in Table 1 and *k* is the rate constant which depend on the temperature after the Arrhenius law

$$k = A\exp(-E_a / RT) \tag{3}$$

where A is the Arrhenius preexponential factor, E_a is the activation energy, T is the absolute temperature and *R* is the absolute gas constant.

The Eq. (4), obtained by substituting Eq. (3) into Eq. (1), is generally used for performing the kinetic analysis of non-isothermal data

$$\frac{d\alpha}{dt} = A\exp(-E_a / RT)f(\alpha) \tag{4}$$

However, MacCallum and Tanner [1] questioned the correctness of Eq. (4)] and an extensive debate started in which number of authors [2–9] supported the MacCallum and Tanner's arguments. These authors considered that since α is a function of both temperature and time

$$\alpha = f(T,t) \tag{5}$$

the total differential is given by the following constitutive equation

$$\frac{d\alpha}{dt} = \left(\frac{\partial\alpha}{\partial t}\right)_T + \left(\frac{\partial\alpha}{\partial T}\right)_t \left(\frac{\partial T}{\partial t}\right)_\alpha \tag{6}$$

Some authors [2,4,5] proposed a philosophical solution to Eq. (6) by reasoning that no change in α can occur at a constant time, that led them to assume that $(\partial\alpha/\partial T)_t = 0$ and then

$$\frac{d\alpha}{dt} = \left(\frac{\partial\alpha}{\partial t}\right)_T \tag{7}$$

Table 1. f(α) and g(α) kinetic functions

Mechanism	[1]Symbol	f(α)	g(α)
Phase boundary controlled reaction (contracting area)	R2	$(1-\alpha)^{1/2}$	$2\left[1-(1-\alpha)^{1/2}\right]$
Phase boundary controlled reaction (contracting volume)	R3	$(1-\alpha)^{2/3}$	$3\left[1-(1-\alpha)^{1/3}\right]$
Random nucleation followed by an instantaneous growth of nuclei. (Avrami-Erofeev Eq. n =1)	[2]F1	$(1-\alpha)$	$-\ln(1-\alpha)$
Random nucleation and growth of nuclei through different nucleation and nucleus growth models. (Avrami-Erofeev equation.)	An	$n(1-\alpha)[-\ln(1-\alpha)]^{1-1/n}$	$[-\ln(1-\alpha)]^{1/n}$
Two-dimensional diffusion	D2	$1/[-\ln(1-\alpha)]$	$(1-\alpha)\ln(1-\alpha)+\alpha$
Three-dimensional diffusion (Jander equation)	D3	$\dfrac{3(1-\alpha)^{2/3}}{2\left[1-(1-\alpha)^{1/3}\right]}$	$\left[1-(1-\alpha)^{1/3}\right]^2$
Three-dimensional diffusion (Ginstling-Brounshtein equation)	D4	$\dfrac{3}{2\left[(1-\alpha)^{-1/3}-1\right]}$	$(1-2\alpha/3)-(1-\alpha)^{2/3}$

[1]The symbols of Sharp and Wentworth (J.H. Sharp and S.A. Wentworth, Anal. Chem., 41 (1969) 2060) have been used.

[2]This equation represents an Avrami-Erofeev kinetic model with n=1 instead of a first order reaction as assumed by Sharp and Wentworth. The symbol A1 would be more proper.

Thus, according to these authors [1–5], Eq. (4) could be used for the kinetic analysis of non-isothermal data, provided that the partial derivative with regards to the time at constant temperature would be identical to the total derivative. On the other hand, McCallum and Tanner [1,3,11] and others [6–10] considered that Eq. (7) is valid only when $(\partial T/\partial t)_\alpha = 0$ (i.e., under isothermal conditions). They proposed for $(\partial\alpha/\partial T)_t$ the following relationship, as obtained from Eqs. (2) and (3)

$$\left(\frac{\partial\alpha}{\partial T}\right)_t = \frac{AE_a}{RT^2}\exp\ (-E_a/RT)\ f(\alpha)t \tag{8}$$

and considered that unless the function $(\partial T/\partial t)_\alpha$ was previously known, the total derivative given by Eq. (6) could not be evaluated. McCallum and Tanner [1], Norwisz [6], Dutta and Ryan [7] and Blazejowsky [8] (after assuming a constant heating rate $\beta = (T - T_0)/t$, T_0 being the starting temperature), obtained from Eqs. (3), (6), (7) and (8) the following expression:

$$\frac{d\alpha}{dt} = A\exp(-E_a/RT)\left[1+\frac{E_a}{RT}\,\frac{(T\;-\;T_0)}{T}\right]f(\alpha) \tag{9}$$

These authors considered Eq. (9) instead of Eq. (4) as the correct expression of the reaction rate under a constant heating rate. However, Koch et al. [12] carried out a comparative study of the applicability of Eqs. (4) and (9) by evaluating more than 90 homogeneous organic reactions in the liquid phase, and they observed that only the kinetic parameters calculated from Eq. (4) showed a very good agreement with those determined from isothermal runs. Therefore, these authors concluded that Eq. (9) does not seems to be justified, but no theoretical support for their finding was reported. Moreover, Sesták [13] reviewed the crystallisation kinetics of some metallic glasses studied by complementary isothermal and non-isothermal methods and reported an excellent agreement between the activation energies determined from both methods when Eq. (4) was assumed to be correct. A very good agreement between the kinetic parameters obtained from isothermal and non-isothermal data obtained from Eq. (4) for the thermal decomposition of inorganic compounds were also reported [14–23]. On the other hand, Brown, Dollimore and Galwey [24] remarked that for some reversible thermal decomposition reactions, the agreement is less than satisfactory. Nevertheless, this behaviour could be attributed to the influence of the transport of gases that would depend on the experimental conditions used, and it does not necessarily support the inconsistency of Eq. (4) for being applied to the kinetic analysis of nonisothermal data. In fact Garn [25] pointed out that the influence of heat and mass transfer phenomena on the rate of reversible reactions is higher under rising temperature experiments than under isothermal conditions. Galwey and Brown [26,27], in recent reviews on the topic, proposed the use of Eq. (4) for the kinetic analysis of non-isothermal data although they remarked that the correctness of this basic constitutive equation has aroused a debate that still continues since its physical meaning was questioned by McCallum and Tanner. For instance, they reported a very strong criticism of Boldyreva [28] who concluded that the widely spread use of non isothermal methods stands on the rapidity with which the information is obtained which “may compensate for the absence of physical meaning”.

The MacCallum and Tanner arguments were refuted by a number of authors [29–34] that pointed out that the derivation of Eq. (6) from Eq. (5) has meaning only if the two variable considered (T and t) are independent variables, what is not true in the case of linear rising temperature experiments in which it is quite clear that the temperature is a predefined function of the time. This fact would explain the failure of Eq. (9) for obtaining values of the activation energy consistent with those obtained for the same process from isothermal experiments.

It is noteworthy to remark that the philosophical arguments that led to assume that the term $(\partial\alpha/\partial T)_t$ is cancelled under isothermal conditions seems to be far from being supported by mathematics. The term of $(\partial\alpha/\partial T)_t$ must represent the derivative of the α-T plot obtained from a set of isotherms at a fixed reaction time. In order to test this interpretation a series of isothermal curves were calculated [32] from Eq. (2) at intervals of 1K by assuming a F1 kinetic model and the following kinetic parameters: $E_a = 100$ kJ/mol and $A = 10^4\ s^{-1}$. The values of α taken from these isothermal curves at a preset time $t = 600$ s have been represented in Figure 4-1 against the temperatures together with the plot of $(\partial\alpha/\partial T)_t$ versus the temperature calculated from this α-T by numerical differentiation. If the above interpretation on the $(\partial\alpha/\partial T)_t$ meaning were correct it would be expected that (according with the logarithmic expression of Eq. (8) the plot of $[\ln (\partial\alpha/\partial T)_t + 2 \ln T - \ln f(\alpha)]$ against 1/T would lead to a straight line whose slope and intercept would give the activation energy and the Arrhenius pre-exponential factor, respectively. The values of $[\ln (\partial\alpha/\partial T)_t + 2 \ln T - \ln f(\alpha)]$ calculated from the $(\partial\alpha/\partial T)_t$ values taken from Figure 4-1 have been plotted against 1/T in Figure 4-2. The kinetic parameters obtained from this plot ($E = 100$ kJ/mol and $A = 10^4\ s^{-1}$) agree quite well with the values previously assumed for constructing the isothermals used for calculating the plots included in Figure 4-1, confirming the meaning that we have previously assessed to $(\partial\alpha/\partial T)_t$.

The above analysis allows us to conclude that Eq. (9) proposed for representing the expression of the reaction rate under linear heating rate conditions was badly defined. This is because Eqs. (5) and (6) are only representatives of isothermal experiments, provided that only in this case temperature and time are independent variables. It is clear that this conclusion would be extended to any other kinetic expression that would be developed from Eqs. (5) and (6). One would conclude that the controversy still alive about the usefulness of Eq. (4) for the kinetic analysis of non-isothermal data has not any physical support. Sesták [32] summarizes this point in the following way: “incorrect use of function $\alpha = f(T, t)$ as a constitutional equation is a typical example of where purely mathematical treatment of a problem that requires logical analysis can lead”.

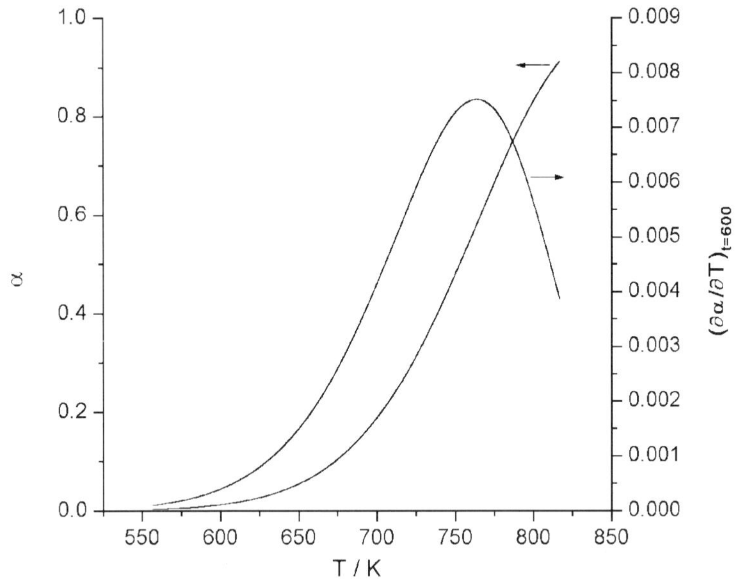

Figure 4-1. Values of α and $(\partial\alpha/\partial T)_t$ at a time t = 600 s as a function of the temperature of the corresponding isothermal runs.

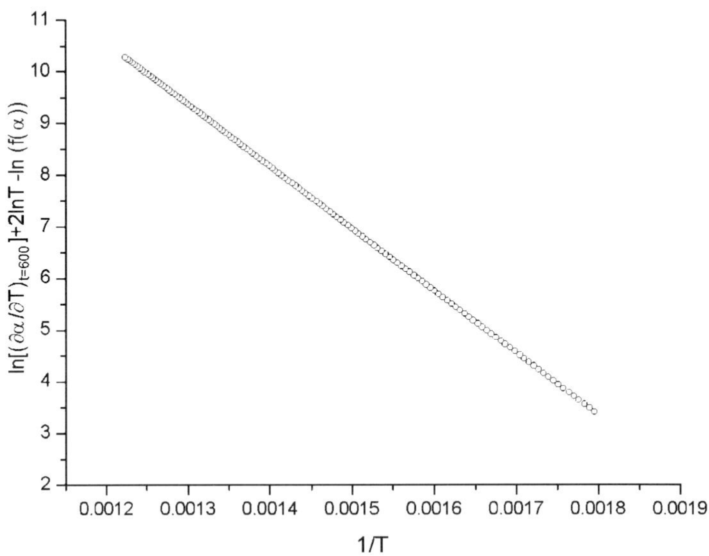

Figure 4-2. Plot of the values of $\ln[(\partial\alpha/\partial T)_{t=600}] + 2\ln T - \ln(f(\alpha))$ (calculated from data in Figure 4-1) versus 1/T.

Sesták [35] demonstrated that Eq. (4) works with linear heating experimental data departing from a constitutive equation in which the time and the heating rate were the independent variables chosen. The Sestak methodology could be extended to any other non isothermal method by bearing in mind that a proper constitutive equation would be a function of the time and other independent variable, ϕ, representative of the experimental method to be considered.

$$\alpha = f(t,\ \phi) \tag{10}$$

ϕ could be the heating rate in the case of rising temperature experiment as considered by Sestak [35] but, in the case of Sample Controlled Thermal Analysis (SCTA), it could also represent either the constant reaction rate or the constant acceleration of the rate or any function of the reaction rate independent of the time.

Provided that the variables t and ϕ are independent, one gets for the total derivative of α with respect to time

$$\frac{d\alpha}{dt} = \left(\frac{\partial\alpha}{\partial t}\right)_{\phi} + \left(\frac{\partial\alpha}{\partial\phi}\right)_{t}\left(\frac{\partial\phi}{\partial t}\right)_{\alpha} \tag{11}$$

It is quite clear that $(\partial\phi/\partial t)_{\alpha} = 0$ for a particular experiments controlled by the variable ϕ and Eq. (11) becomes

$$\frac{d\alpha}{dt} = \left(\frac{\partial\alpha}{\partial t}\right)_{\phi} \tag{12}$$

Equation (12) is similar to Eq. (7) and demonstrates that the reaction rate is identical to the total derivative given by Eq. (4) whatever would be the thermal pathway followed by the reaction. In other words, one can say, acording to Sesták [13], that "the reacted fraction is not a state function of temperature and time ($\alpha \neq f(T, t)$ like the derivative $d\alpha/dt$. Therefore, any additional term of the type $(d\alpha/dT)_t$ is unjustified and thus invalid."

4.4. Advantages of SCTA

The Sample Controlled Thermal Analysis (SCTA) methods have important advantages for performing the kinetic analysis of solid state reactions with regard to the conventional non-isothermal and even isothermal methods. This feature is based on the following two statements which will be discussed later on in this chapter:

1. The precise control of the reaction rate involved in SCTA implies a control either direct or indirect of both the partial pressure of the gases generated or consumed in the reaction and the associated heat evolution rate. This control allows to minimize the heat and mass transfer phenomena and to obtain real kinetic parameters of the forward reaction far from the equilibrium conditions.
2. The resolution power of SCTA for discriminating among the reaction kinetic models of solid state reactions is somewhat more favourable than that of isothermal methods but considerably higher than that of conventional non-isothermal methods.

The graphic representation proposed by Reading [36,37], shown in Figure 4-3, is very useful for remarking the advantages of SCTA with regards to conventional methods from the point of view of the experimental condition control. Figure 4-3 represents the evolution of the temperature, mass change and partial pressure of the evolved gases as a function of the time. In general, the lower the reaction rate, the lower is the chance of appreciable temperature or pressure gradients in the sample bed. The advantage of Constant Rate Thermal Analysis (CRTA) in terms of maintaining constant the product gas pressure and

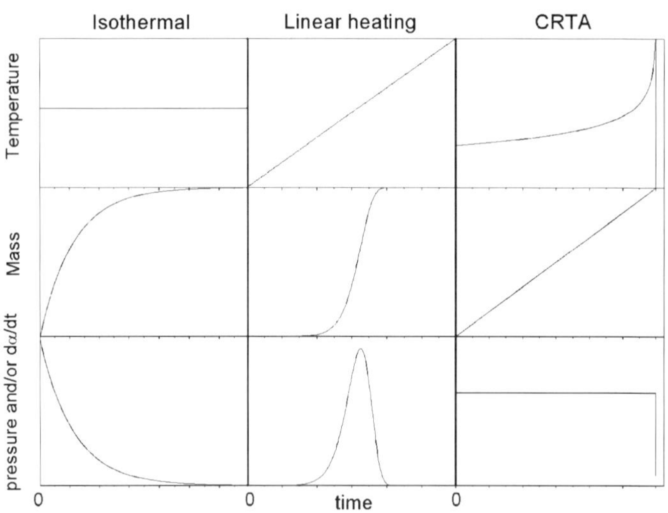

Figure 4-3. Trend of temperature, mass and pressure (and/or dα/dt) during a thermal decomposition reaction for different methods.

the reaction rate at a strictly constant value becomes clear from Figure 4-3. Thus, SCTA method allows to minimize the pressure and temperature gradients within the sample and, therefore, to minimize or even avoiding the influence of heat and mass transfer phenomena on the forward reaction, leading to meaningful kinetic parameters from an adequated kinetic analysis. On the other hand, the isothermal and conventional rising temperature methods (Fig. 4-3) would lead to significant changes in the reaction rate and in the product gas pressure, that generally cannot be controlled by the user and could modify the shape of the α-T plots leading to a meaningless interpretation of the reaction mechanism. The good control that SCTA methods exert on both the atmosphere surrounding the sample and the real temperature of the sample bed explains that it were frequently observed that the activation energies calculated for either reversible [38–52] or irreversible [53] reactions of thermal decomposition of solids were independent of the sample size in a wide range of starting sample weight, while a similar behaviour was not observed when rising temperature experiments were concerned. Thus, Criado et al. [38] reported that the activation energy for the thermal decomposition of calcium carbonate, as determined under a high dynamic vacuum from SCTA experiments, was independent of the starting sample weight in the investigated range, i.e. from 0.5 mg to 50 mg. However, the activation energy obtained for this compound under high vacuum from conventional linear heating TG diagrams is strongly depending on the experimental conditions and it is necessary to decrease the sample weight below 2 mg and to use an heating rate lower than 1 K/min in order to obtain kinetic parameters independent of the selected experimental conditions. These results are consistent with those later on reported by Reading et al. [42] for the same reaction. In the case of the thermal decomposition of $CdCO_3$ and $PbCO_3$ it was not even possible to get activation energy values independent of sample size and heating rate from experiments at rising temperature, while the activation energies obtained from SCTA were independent of the experimental conditions in a wide range of sample sizes [41]. Ortega et al. [44,49] reached a similar conclusion studying the kinetics of the thermal decomposition of dolomite. Thus, it can be concluded that SCTA is the most reliable approach for obtaining meaningful kinetic parameters for thermal decomposition reactions.

The influence of the pressure of the gas product around the sample was analysed by several authors [47,54–56] and it was concluded that, in the case of reversible reaction, a poor control of the pressure would lead to artificially high values of the activation energy, although reliable kinetic parameters of the forward reaction would be obtained by introducing a pressure accommodation function [55] in the general kinetic equation, provided that the pressure of the product gas around the sample is known. On the other hand, Rouquerol et al. [57,58] proved that even a relatively small change of the product gas pressure in the high vacuum range could modify the reaction mechanism of the dehydration of inorganic hydrates. This behaviour would explain that the activation energies obtained from SCTA methods are independent of the sample size while the values of the kinetic parameters obtained from rising temperature methods are very often dependent on the sample size and the heating rate. This is because all the SCTA kinetic analyses of reversible reactions referred in the previous paragraph were carried out from experimental data obtained under vacuum using experimental tools based on the method formerly developed by Rouquerol [59,60]. This method implies to control the reaction temperature by controlling the residual pressure in such a way that it is maintained constant at a very low value previously selected by the user. Thus, the influence of the mass transfer phenomenon is minimized. Controlling the pressure from linear heating rate experiments is rather difficult, as illustrated by Criado et al. [38] from a set of TG diagrams of $CaCO_3$ recorded under a dynamic starting vacuum of 2.6×10^{-6} mbar using a high pumping rate vacuum system. The pressure and the weight change were recorded simultaneously and it was reported that for a initial sample weight of 21 mg, the pressure increased up to about 10^{-3} mbar at a heating rate of 10 K/min. It was necessary to decrease the sample weight to 1 mg and the heating rate to 0.5 K/min for avoiding the pressure to exceed the starting value of 2.6×10^{-6} mbar. However, there was no problem for recording SCTA experiments within a wide range of sample weights and at a constant pressure as low as 5×10^{-6} mbar using the same experimental tool. Moreover, it has been recently reported [61] the kinetic analysis of the thermal decomposition of $BaCO_3$ from SCTA experiments under constant residual pressures of CO_2 lower than 10^{-7} mbar by using a high vacuum system equipped with a mass spectrometer attached to an thermobalance. In this case, the partial pressure of the gas released in the reaction is directly monitored during the entire experiment by means of the mass spectrometer and even partial pressures lower than the total limit vacuum of the system are used for the control. By means of this procedure, the forward reaction for the thermal decompositions of very stable compounds with low equilibrium pressures could be studied in conditions far from equilibrium.

The above results supports our first statement and allow us to conclude that SCTA methods are a more reliable approach than conventional non-isothermal methods in order to obtain correct values of the activation energy of thermal decomposition of solids. The power of SCTA for discriminating the kinetic model of the reaction will be analysed in the next section. This analysis will be mainly based on the Constant Rate Thermal Analysis (CRTA) method that has been the most extensively and systematically SCTA method used in literature, as shown in previous reviews on the topic [36,37,52].

4.5. Kinetic Analysis of CRTA Curves

If the α-T (or t) plot is obtained at a constant decomposition rate ($C = d\alpha/dt$), Eq. (4) can be rearranged, after taking logarithms, in the form

$$-\ln f(\alpha) = \ln\frac{A}{C} - \frac{E_a}{RT} \tag{13}$$

The plot of the left hand side of Eq. (13) as a function of 1/T lead to a straight line where the slope leads to the activation energy and the intercept to the pre-exponential factor of the Arrhenius expression of the process, provided that the proper f(α) function were selected, except if the kinetic model were represented by the function $f(\alpha) = (1 - \alpha)^n$ (i.e. R2, R3 and F1 models, frequently referred as "*n* order" reactions). In such a case, Eq. (13) becomes:

$$\ln\frac{1}{(1-\alpha)} = \ln\frac{A}{nC} - \frac{E_a}{nRT} \tag{14}$$

and E_a and n cannot be simultaneously determined from a single experiment [62–63] unless one of these two parameters were known from other source. The first comparative study between conventional linear heating and CRTA methods with regards to their ability for discriminating among the set of kinetic models described in Table (1) was carried out by Criado [63]. This work supported the hypothesis that a single CRTA curve allows to discern among Avrami-Erofeev, "*n* order" and diffusion controlled reactions; this discrimination is not possible from a single curve obtained under a linear heating program, as firstly outlined by Criado et al. [64,65] and later on confirmed by other authors [66–74]. A theoretical analysis of the correlation between the activation energies obtained from the kinetic analysis of a single TG curve and the kinetic model previously assumed for determining the kinetic parameters was carried out by Criado et al. [66] and Málek [72]. The shape analysis of the CRTA curves as a function of the kinetic model would be very illustrative for understanding the power of SCTA for discriminating the real reaction kinetic model.

4.6. The Shape of CRTA Curves

The interest of analysing the shape of this curves by determining the α values at which maxima, minima or inflection points appears on the temperature versus α plots calculated for the f(α) functions shown in Table 1 has been shown in a previous paper [70].

Table 2. f(α) kinetic functions and their first and second derivatives

Symbol	f(α)	$f'(\alpha)$	$f''(\alpha)$
R2	$(1-\alpha)^{1/2}$	$\dfrac{-1}{2(1-\alpha)^{1/2}}$	$\dfrac{-1}{4(1-\alpha)^{3/2}}$
R3	$(1-\alpha)^{2/3}$	$\dfrac{-2}{3(1-\alpha)^{1/3}}$	$\dfrac{-2}{9(1-\alpha)^{4/3}}$
F1	$(1-\alpha)$	-1	0
An	$n(1-\alpha)[-\ln(1-\alpha)]^{1-1/n}$	$\dfrac{n\ln(1-\alpha)+n-1}{[-\ln(1-\alpha)]^{1/n}}$	$\dfrac{-n+1+\dfrac{1-1/n}{\ln(1-\alpha)}}{(1-\alpha)[-\ln(1-\alpha)]^{1/n}}$
D2	$1/[-\ln(1-\alpha)]$	$\dfrac{-1}{[\ln(1-\alpha)]^2(1-\alpha)}$	$\dfrac{-2-\ln(1-\alpha)}{(1-\alpha)^2[\ln(1-\alpha)]^3}$
D3	$\dfrac{3\ (1\ -\ \alpha)^{2/3}}{2\ \left[1-(1\ -\ \alpha)^{1/3}\right]}$	$\dfrac{1/2-(1-\alpha)^{-1/3}}{\left[1-(1-\alpha)^{1/3}\right]^2}$	$\dfrac{3(1-\alpha)^{-1/3}-(1-\alpha)^{-2/3}-1}{3(1-\alpha)^{2/3}\left[1-(1-\alpha)^{1/3}\right]^3}$
D4	$\dfrac{3}{2\left[(1-\alpha)^{-1/3}-1\right]}$	$\dfrac{-(1-\alpha)^{-4/3}}{2\left[(1-\alpha)^{-1/3}-1\right]^2}$	$\dfrac{2\left[(1-\alpha)^{1/3}-1\right]+1}{3(1-\alpha)^{8/3}\left[(1-\alpha)^{-1/3}-1\right]^3}$

Equations (15) and (16) represent the first and second derivatives of T with respect to α, as obtained from Eq. (13)

$$\frac{dT}{d\alpha} = -\ \frac{RT^2}{E_a}\ \frac{f'(\alpha)}{f(\alpha)} \tag{15}$$

$$\frac{d^2\ T}{d\alpha^2} = \frac{RT^2}{E_a}\left[\frac{2RT}{E_a}\left(\frac{f'(\alpha}{f(\alpha)}\right)^2 - \frac{f''(\alpha)\ f(\alpha)\ -\ f'(\alpha)^2}{f(\alpha)^2}\right] \tag{16}$$

The f′(α) and f″(α) functions are listed in Table 2 together with the corresponding f(α) functions.

A plot of T as a function of α show a maximum or a minimum at the α_m value at which $dT/d\alpha = 0$, i.e.

$$f'(\alpha_m) = 0 \tag{17}$$

Solutions to Eq. (17) have been found only for Avrami-Erofeev kinetic models with $n > 1$:

$$\alpha_m = 1 - \exp\left(\frac{1-n}{n}\right) \tag{18}$$

Equation (18) points out that α_m values are independent of E/RT leading to a value of $d^2T/d\alpha^2 > 0$ at α_m, indicating that the T versus α CRTA plots of reactions fitting an Avrami-Erofeev law yield a minimum at the reacted fraction α_m.

On the other hand, it would be expected that the T versus α plots show inflection points where α_i give $d^2T/d\alpha^2 = 0$. Therefore, according to Eq. (16) α_i must fulfill the following condition

$$[f'(\alpha_i)]\left[\frac{RT}{2E_a} + 1\right] - f(\alpha_i)\ f''(\alpha_i) = 0 \tag{19}$$

It was found that only diffusion-controlled reactions lead to a solution of Eq. (19). The α_i expressions obtained for the kinetic models D2, D3 and D4 and the corresponding values of α_i obtained as a function of E_a/RT are given in reference [70]. In the case of "*n* order reactions" (R2, R3 and F1) neither maxima nor minima nor inflection points were found [70] but was concluded from the analysis of the Eq. (4) that the plot of α as a function of temperature is convex all over the α range.

The above analysis suggest that the shape of the CRTA curves is strongly dependent on the reaction mechanism. This conclusion was confirmed from the shape analysis of the set of α versus T plots included in Figure 4-4 that were simulated for the different kinetic models described in Table (1). It is clearly shown that the curves corresponding to "*n* order" models are convex with regards to the T axis, while the corresponding ones to diffusion kinetic equations present an inflection point and, finally, those coming from the Avrami-Erofeev equation show a minimum with regard to the α axis. This behaviour forces to the α-T profiles of reactions following the Avrami-Erofeev mechanism to start with a rise in temperature until reaching the preset value of the reaction rate. This step is immediately followed by a temperature fall until reaching a determined value of the reacted fraction, α_m, that depends on the value of the coefficient *n* of the Avrami-Erofeev equation, and again the temperature rises once the corresponding α_m value is attained ($\alpha_m = 0.393$; 0.486

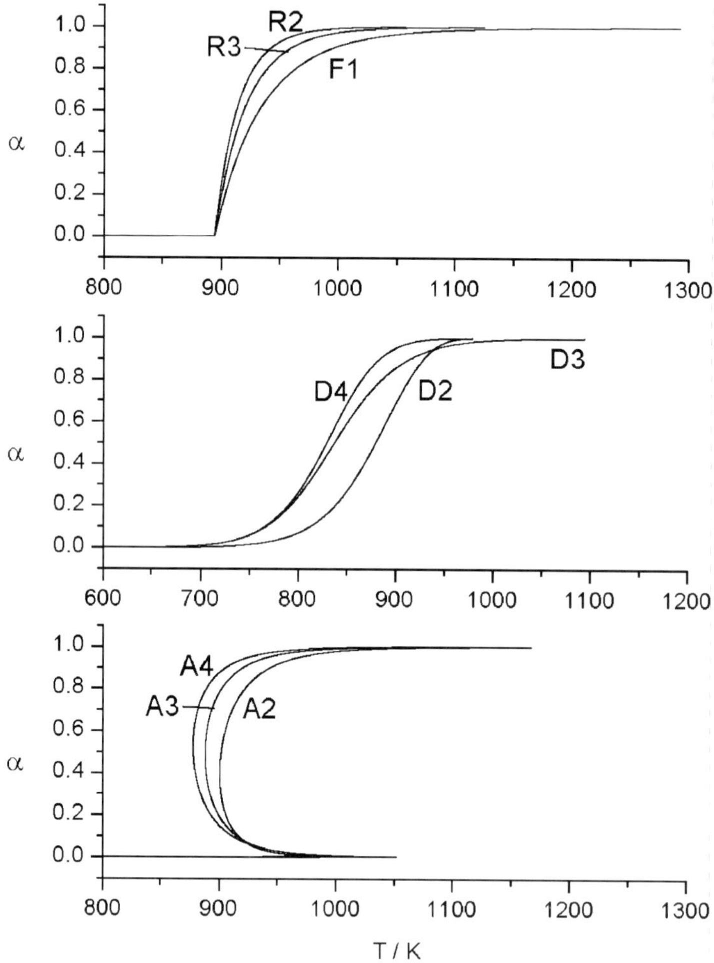

Figure 4-4. Shapes of CRTA curves corresponding to different kinetic models simulated assuming $E_a = 167$ kJ/mol, $A = 1.7 \times 10^6\ s^{-1}$ and $C = 3 \times 10^{-4}\ s^{-1}$.

and 0.528 for the models A2, A3 and A4, respectively) [70]. It is noteworthy to point out that the α_m values calculated for CRTA curves exactly agree with the values previously calculated by Criado [75] for the reacted fraction at which the reaction rate reach a maximum under isothermal conditions that marks the end of the acceleratory period [76]. This finding supports the proposal of Barnes et al. [77] that consider that the part of the α-T profile on which the curve back on itself would correspond to the acceleratory period in which the total surface area of the growing nuclei would increase leading to an acceleration of the reaction that would be offset by a diminution of the temperature. The later rising temperature stage would correspond to the decay period on which the growing nuclei overlap producing a decrease of the interface reaction area that would lead to a decrease of the reaction rate that must be compensated by a rise in temperature in order to maintain constant the reaction rate. The thermal decomposition of anhydrous nickel nitrate would be a convenient reaction for an experimental check of the peculiar shape of the CRTA plots of reaction fitting an Avrami-Erofeev equation. This transformation was previously studied [78] under isothermal conditions and it was concluded that it obeys the Avrami-Erofeev kinetic law with an exponent $n = 2$. Figure 4-5 shows an CRTA α-T plot obtained [70] for the thermal decomposition of the same sample of anhydrous nickel nitrate. The shape of this plot is identical to that forecasted for an A2 Avrami-Eroffev model with a minimum on the α axis at $\alpha = 0.38$. CRTA plots in which the temperature back on itself have been very often reported in literature for different reactions [57,58,79–89]. The results reported

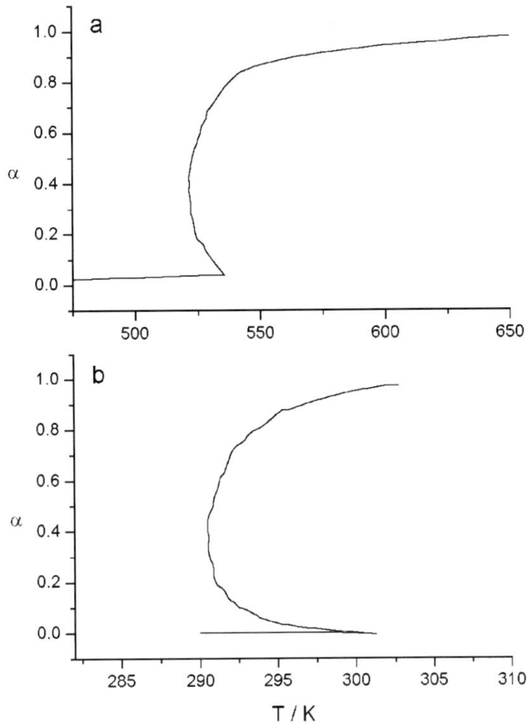

Figure 4-5. CRTA Curve for a) the thermal decomposition of anhydrous nickel nitrate under vacuum at C = (1) 7×10^{-5} s^{-1}, and b) dehydration of uranyl nitrate trihydrate under vacuum at C = 2.77×10^{-6} s^{-1}.

by Bordère et al. [58] for the dehydration of uranyl nitrate trihydrate constitutes other nice example of a typical plot for an A2 kinetic model as illustrated in Figure 4-5.

To summarize, we can conclude that the CRTA method allows to discriminate among phase boundary, Avrami-Erofeev and diffusion controlled processes because the difference in the shape of their corresponding α-T plots are so dramatic that a mere glance at the shape provides and easy way of discriminating between these three groups of kinetic models. This rapid discrimination of the kinetic model is less favourable for isothermal methods because in such a case the α-t plots of the reaction fitting an Avrami-Eroffeev equation are sigmoid-shaped but these plots are convex with regards to the t axis for the phase boundary and diffusion controlled reactions [70] and they cannot be discriminated without an ulterior kinetic analysis. Finally, the discrimination renders quite impossible if α-T curves are obtained under a linear heating program because in such a case sigmoid-shaped curves are obtained for any reaction mechanism. In order to illustrate this behaviour it has been shown in Figure 4-6 that a unique TG curve can be calculated by assuming different kinetic models: F1, A2 and A3. However, these models can be unambiguously discriminated by CRTA as shown in Figure 4-7 that represents the curves calculated from Eq. (4) for the models F1, A2 and A3 by assuming the same kinetic parameters used in Figure 4-6 and a constant reaction rate $C = 5 \times 10^{-6}\ s^{-1}$.

4.7. Master Plots

A master curve is a graphic representation of a dimensionless parameter (that is directly obtained from the raw experimental data) as a function of the reacted fraction. These plots behave like a portrait of the reaction models and can be used for their identification. The "reduced time" master curves [90–94] are very commonly used for discriminating the kinetic model from a single α-t isothermal curve, while the master "reduced rate plots" have been proposed for the kinetic analysis of isothermal $(d\alpha/dt)$-t traces [92,94,95]. Different series of master curves have been also proposed for discerning the kinetic model from TG or DSC obtained under linear heating [94–105], but in this case a previous determination of the activation energy from isoconversional methods is required for discerning the kinetic model. The first proposal of master plots for discriminating the kinetic equation fitted by solid state reactions from CRTA data is due to Reading [106]. This author developed the so called "reduced temperature" master plots. The reduced temperature was calculated taking the temperature $T_{0.9}$ at which $\alpha = 0.9$ and the temperature $T_{0.3}$ at which $\alpha = 0.3$ as

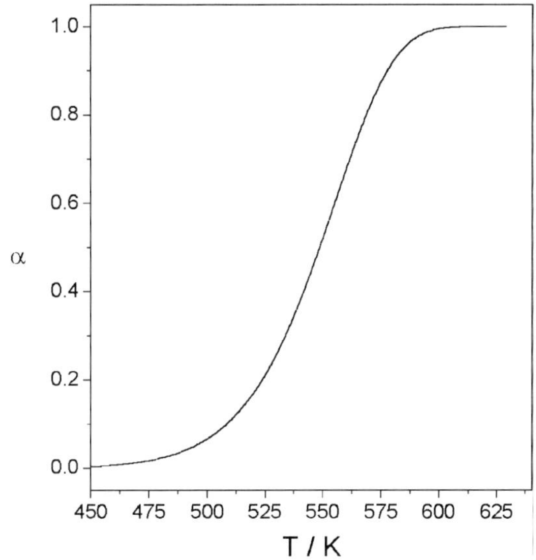

Figure 4-6. A single TG curve at a heating rate of 1ºC/min for three different models: F1: E_a = 172.3 kJ mol^{-1} and A = 2.30 × 10^{13} s^{-1}; A2: E_a = 118.1 kJ mol^{-1} and A = 1.24 × 10^{8} s^{-1}; A3: E_a = 100.0 kJ mol^{-1} and A = 1.66 × 10^{6} s^{-1}.

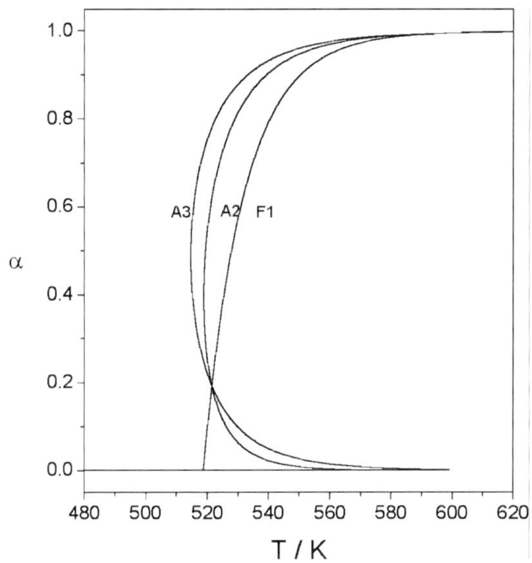

Figure 4-7. Discrimination between the three kinetic models (F1, A2, A3) of Figure 4-6 by means of the CRTA method.

reference points in the CRTA curve. Thus, it can be easily obtained from Eq. (14) which gives the following expression

$$\frac{\dfrac{1}{T_\alpha} - \dfrac{1}{T_{0.9}}}{\dfrac{1}{T_{0.3}} - \dfrac{1}{T_{0.9}}} = \frac{\ln f(\alpha) - \ln f(0.9)}{\ln f(0.3) - \ln f(0.9)} \tag{20}$$

The left hand side of Eq. (20), which would be directly calculated at given values of α, was called by Reading "reduced temperature". Eq. (20) indicates that, at a given α, the experimentally determined value of the reduced temperature and the theoretically calculated value of [ln f(α) – ln f(0.9)]/[ln f(0.3) – ln f(0.9)] are equivalent when an appropriate f(α) for describing the rate process under investigation is applied. Because both the values depend only on α, the plot of the right hand side of Eq. (20) against α would represent the master plots of "reduced temperature" as a function of the reacted fraction. The master plots obtained from Eq. (20) are included in Figure 4-8[1]. It is noteworthy to point out that although it is not possible to distinguish the "reaction order" from these

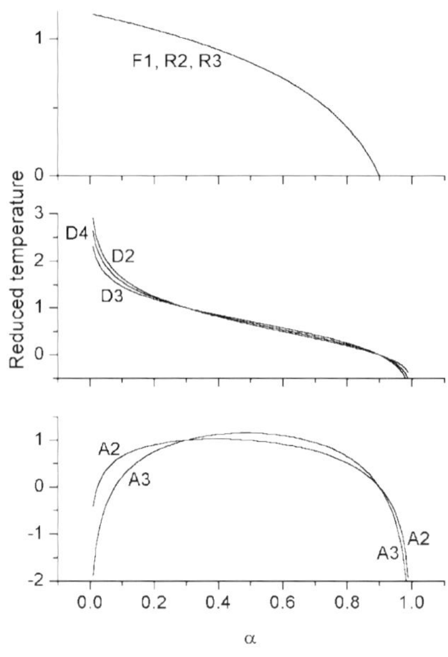

Figure 4-8. Reduced temperature master plots.

[1]The master plots shown in Figure 8 has been recalculated here because some errors were squeezed in the original paper [106].

master curves, they provide a means of distinguishing among all the other kinetic models with more sensitivity than conventional techniques [37], in agreement with the conclusions drawn in Section 5.

Pérez-Maqueda et al. [107] proposed a series of "reduced rate" master plots departing from the first derivative of Eq. (13) with regards to T which, after taking into account the identity dT/dt = C/(dα/dT) under CRTA conditions, can be rearranged in the form

$$T^2 \frac{d\alpha}{dT} = \frac{-E_a f(\alpha)}{R f'(\alpha)} \tag{21}$$

The "reduced rates" were calculated [107] by taking as reference point the absolute temperature and rate at which $\alpha = 0.5$, using following expression

$$\left(\frac{T}{T_{0.5}}\right)^2 \frac{(d\alpha / dT)}{(d\alpha / dT)_{0.5}} = \frac{f'(0.5)}{f(0.5)} \frac{f(\alpha)}{f'(\alpha)} \tag{22}$$

where the index 0.5 refers to the values of the quoted parameters or functions at $\alpha = 0.5$.

The left hand side of Eq. (22) was named "reduced rate". This equation indicates that the plot of the "reduced rate" versus α depends neither on the kinetic parameters nor on the preset reaction rate C, but only on the reaction mechanism. The master plots calculated from Eq. (22) for the kinetic models quoted in Table 1 are shown in Figure 4-9. It can be observed once again the full overlapping of the "*n* order" reactions master curves in contrast with the good discrimination of all the others kinetic models. Figures 4-8 and 4-9 show that (in spite of the difficulties of determining the activation energy and the "reaction order") just by looking at the shape of these plots one can easily discriminate between the three families of kinetic models (interphase boundary, nucleation and growth of nuclei and diffusion controlled reactions). It should be noted as well that the reduced rate master curves have a considerably higher resolution power for discriminating the exponent *n* of the Avrami-Erofeev equation than the reduced temperature plots. It has been proved that the thermal decomposition of calcium carbonate [38,42,47,108] and nickel nitrate [78] fit an "*n* order" and an Avrami-Erofeev kinetic model with an Avrami-Erofeev exponent $n \approx 2$, respectively. Thus, these reactions were considered suitable [107] for checking the usefulness of these master plots. The good fitting of the experimental data to the corresponding master curves are shown in Figure 4-10.

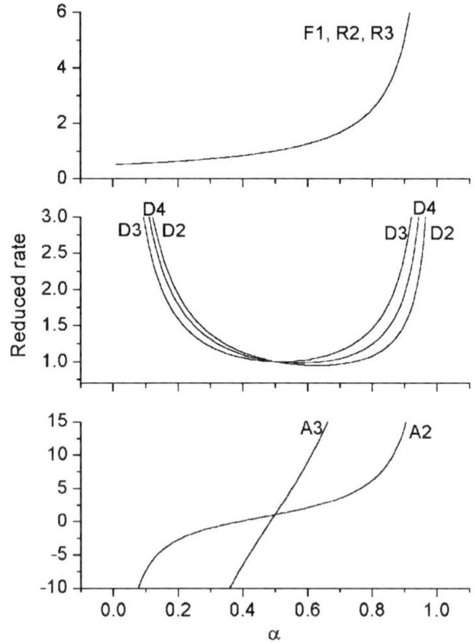

Figure 4-9. Reduced rate master plots.

4.8. Rate-Jump and Related Methods

It has been shown in the previous sections that the kinetic analysis of a single thermoanalytical curve obtained using the Constant Reaction Rate Thermal Analysis method can allow to discriminate without ambiguities among Avrami-Erofeev, "*n* order" and diffusion controlled reactions, but this method is unsuccessful for simultaneously determining the activation energy and the "reaction order" from a single CRTA curve. Rouquerol [109,110] proposed the *Cyclic Reaction Rate* method for overcoming this limitation. This method imposes periodical jumps between two preset decomposition rates at the time that the pressure of the gases generated in the reaction is maintained constant at a previously selected value all over the process. Rouquerol renamed this experimental procedure as *Rate-Jump method* in 1985 [111–113]. This method produces a saw-teeth shaped temperature curve, where each tooth gives assess to an independent calculation of the activation energy, with no assumption about the reaction mechanism. The *Rate-Jump method* compares the state of the sample immediately before the rate jump, at which the reaction rate is C_1 and the temperature is T_1, with the state immediately after the rate jump, at which the reaction rate and the temperature have moved to C_2 and T_2, respectively.

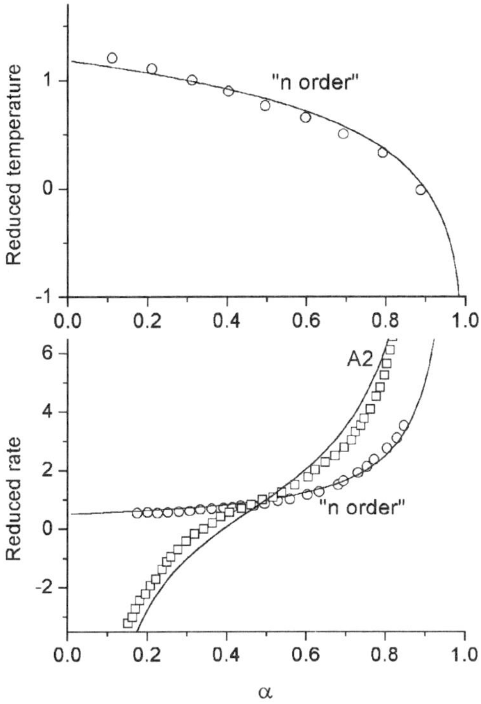

Figure 4-10. A comparison of the experimental reduced temperature and rate plots of the thermal decomposition of $CaCO_3$ (○) and nickel nitrate (□) with the master plots corresponding to "n order" and A2 kinetic model.

By assuming that the two states of the sample to be compared have almost the same reacted fraction, one gets

$$E_a = \frac{RT_1T_2}{(T_2 - T_1)} \ln(C_2/C_1) \tag{23}$$

Equation (23) permits to obtain the activation energy of the process without any assumption regarding to the kinetic law obeyed by the reaction. The value of the "reaction order" n can be determined from Eq. (14) once the activation energy has been determined from Eq. (23). It is advisable to select the two preset reaction rates in such a way that they are sufficiently small for accomplishing that the reacted fraction, α, remains nearly unchanged after the temperature jump from T_1 to T_2, but at the same time the ratio C_1/C_2 should be high enough to assure a large value of $(T_2 - T_1)$ that lead to accurate values of the activation energy. Figure 4-11 shows a detail of one of the teeth taken from a *Rate-Jump* CRTA curve recorded under high vacuum for the thermal

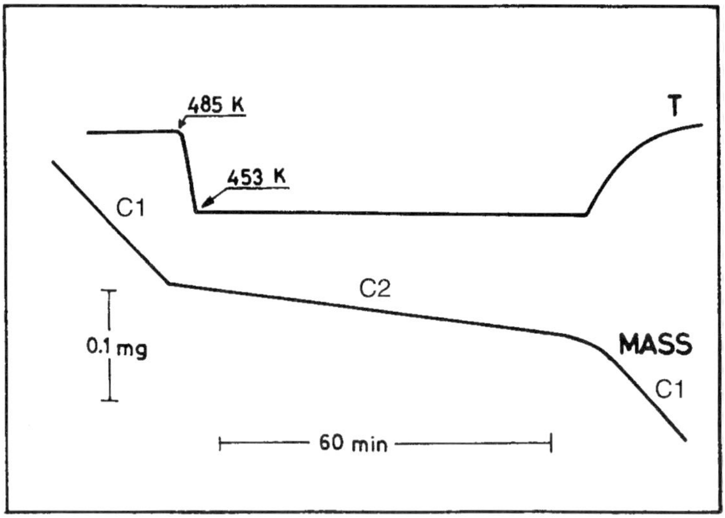

Figure 4-11. A cycle of the CRTA curve of $PbCO_3$ obtained by the rate jump method.

decomposition of $PbCO_3$ [41]. The activation energy calculated from this cycle is E_a = 121 kJ/mol. Very close values of the activation energy were obtained from the 15 cycles recorded in the α range $0.05 < \alpha < 0.95$, leading to a value of (120 ± 10) kJ/mol all over the α range. Moreover, the comparison of this E_a value with the E_a/n value, obtained from the plot of the values calculated for the left hand side of Eq. (14) from a single CRTA curve of $PbCO_3$ as a function of the inverse of the temperature, permitted to conclude that the thermal decomposition of this compound fitted a F1 kinetic model [41]. The combined analysis of *Rate jump* methods with the analysis of a single CRTA curve has been successfully used by many authors for discerning the reaction mechanism of solid state reactions [39–42,49,51,57,58,77,81,88,89,114–125]. Eq. (23) has been also used by some authors [126,127] for determining the activation energy as a function of the reacted fraction from two independent CRTA curves recorded at the constant reaction rates C_1 and C_2, respectively.

A number of authors [42,51] have concluded that the *Rate jump* methods lead to values of the activation energy that are insensitive to the starting sample weight and the rate jump ratio, suggesting according to Reading [36,37], that the value for E_a measured by this method does characterize the fundamental chemical reaction rather than the experimental conditions that are dependent upon heat and mass transfer phenomena. This conclusion support the previous one in Section 3 about the restrain of the influence of heat and mass transfer phenomena in CRTA in comparison with conventional isothermal and non isothermal experiments [41,56,62,128–130].

The *Rate jump* method could also be used to check whether a reaction not fitted by any of the master plots shown in Figures 4-8 and 4-9 is really following an unique mechanism in the whole α range. In such a case an unique E_a value independent of α would be obtained although the kinetic equation representing the mechanism were not among those reported in Table 1. This possibility was pointed out by Reading et al. [37,47], Vyazovkyn et al. [69] and Brown [131], who outlined the following statement: "one of the arguments against the use of discrimination method is that the set of models from which the *best model* is to be chosen is too limited. Hence one of the sets is going to be the *best model* even if the set does not contain the true model". This fact was experimentally corroborated by Málek et al. [105,132–136]. The master plots developed by Criado et al. [137] assuming a previous knowledge of the activation energy calculated from the *Rate jump* method could help to draw a curve representing the real reaction model that would be compared with the master plots representing the "ideal" models shown in Table 1 that are based on simplified geometrical assumptions for the reacting particles, the geometry of the reaction interphase, etc. [105]. Taking as reference point the temperature, $T_{0.5}$, at which the reacted fraction is $\alpha = 0.5$, the following equation is easily obtained from Eq. (13)

$$\ln\frac{f(\alpha)}{f(0.5)} = \frac{E_a}{R}\left[\frac{1}{T_{0.5}} - \frac{1}{T}\right] \tag{24}$$

Provided that $\ln[f(\alpha)/f(0.5)]$ is a function of α, which is equal to $(E_a/R)(1/T_{0.5} - 1/T)$, α can be calculated as a function of the right hand side of Eq. (24). Figure 4-12 shows the master curves developed for the $f(\alpha)$ functions included in Table 1. The plot of the experimental values of $(E_a/R)(1/T_{0.5} - 1/T)$ against α has been used in a recent work [61] for checking the kinetic model of the thermal decomposition of $BaCO_3$ from CRTA experiments recorded under a CO_2 residual pressure of 2.10^{-7} mbar. A mass spectrometer was used for controlling the pressure in order to assure that the reaction took place far from the equilibrium. Figure 4-13 clearly shows the good fitting of the experimental data to the master curve of a F1 kinetic model.

It is worth noting that the α versus $(E_a/R)(1/T_{0.5} - 1/T)$ plots calculated from a set of experimental CRTA curves obtained at different preset reaction rates will necessarily fit the same curve if the reaction studied follows a mechanism represented by an unique function $f(\alpha)$. This is true whatever would be the shape of this function and independently of the fact that this function would be included or not among those describing the "ideal models" assumed for developing the kinetic equation included in Table 1. In summary, we can

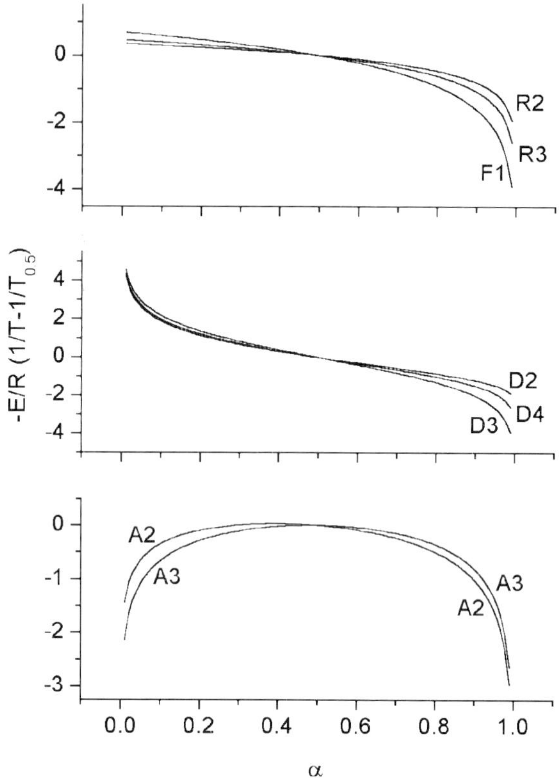

Figure 4-12. Master plots representing $-E_a/RT(1/T - 1/T_{0.5})$ against α.

conclude that combined use of the *Rate jump* method with this kind of master plots is a suitable method for discerning if the reaction kinetics is defined by a unique f(α) function. In such a case both a unique value of the activation energy independent of the α and a unique α versus $(E_a/R)(1/T_{0.5} - 1/T)$ curve will result.

4.9. Relationship between CRTA and other SCTA Methods

It is also of interest to comment the suitability for kinetic analysis of the *quasi-isothermal and quasi-isobaric* (Q-TG) method developed by F. Paulik and J. Paulik [138,139] and the *Stepwise Isothermal Analysis* (SIA) method proposed by Sorensen [140,141]. Other methods more recently developed which can be collected under the umbrella of *dynamic rate* methods [36,37,142–146], are treated in other chapters of this book.

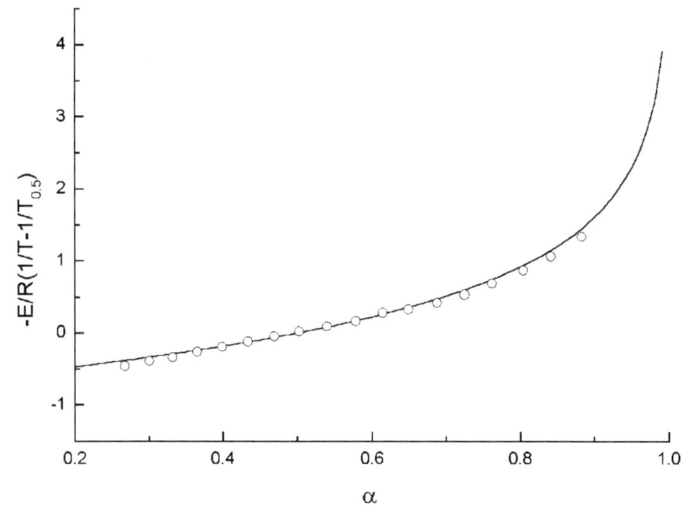

Figure 4-13. A comparison of the kinetic data obtained for the thermal decomposition of $BaCO_3$ (○) with the master plot of an F1 kinetic model (solid line).

The Q-TG method implies to control the temperature in such a way that the rate of the thermal decomposition reaction to be studied remains constant all over the process at a preset value. Thus, CRTA and Q-TG methods are identical from the point of view of the type of function used for controlling the temperature ($d\alpha/dt = C$) and, as a consequence, the method of kinetic analysis outlined above would be applicable, in principle, to the Q-TG method. The main difference between these two methods concerns to the way used for achieving a constant rate and controlling the pressure exerted in the close vicinity of the sample by the gases generated in the reaction: the CRTA method control the reaction rate by maintaining constant a very low pressure while the Q-TG method inevitably lead to high product gas pressure [36,37], generally closed to 1 atmosphere. This fact invalidate the use of the Q-TG method for kinetic analysis [36,37], provided that α-T experimental plots obtained far from the chemical equilibrium is required for obtaining meaningful kinetic parameters. Criado et al. [54] have shown that the experimental conditions imposed by the Q-TG method lead to an α-T plot at a near constant temperature closest to the equilibrium temperature. This behaviour is illustrated in Figure 4-14 where a CRTA and a Q-TG curves obtained for the thermal decomposition of calcium carbonate using samples coming from the same batch [54] are compared. It is observed that under Q-TG conditions the chemical transformation took place at a constant temperature very close to the equilibrium temperature at a CO_2 pressure of 1 bar.

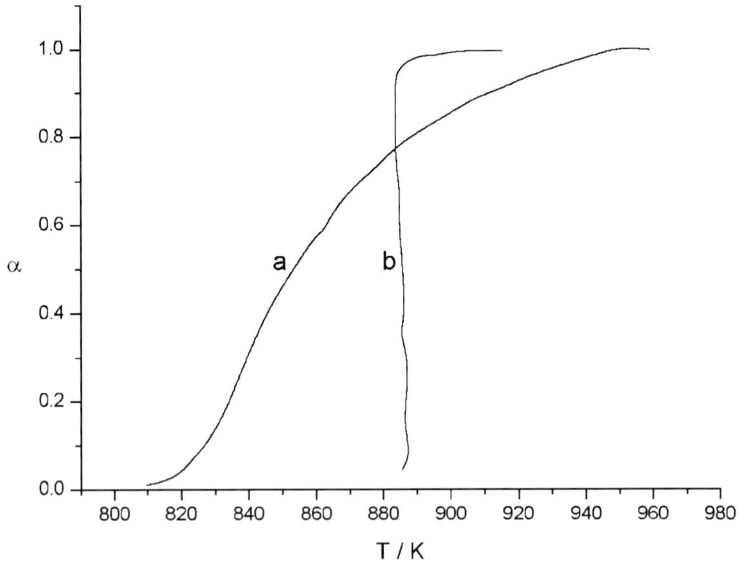

Figure 4-14. CRTA curve (a) and Q-TG curve (b) for the thermal decomposition of $CaCO_3$.

The SIA method imposes to the sample a preset heating rate until the reaction rate exceeds a preset upper limit, C. At this point the increase of temperature stops and the reaction proceeds isothermally until the rate becomes smaller than a preset lower limit, $C - \delta$, when the heating is resumed again. The SIA temperature programme then oscillate between isothermal and rising temperature sections and a single reaction steps could appear as a number of discrete steps which make an accurate discrimination between the different steps in the case of a complex process difficult. In general, it seems to be accepted that SIA approaches to CRTA provided that the upper and lower reaction rate limits are close enough (i.e., $\delta \approx 0$), as shown in Figure 4-15. This feature led Reading [36] to think that "one way of characterising the SIA method is to say that is badly controlled CRTA where the control system fails to provide the smooth continuous response required to maintain the reaction rate at a constant value". However, it is noteworthy to point out that it has been demonstrated from theoretical simulations and experimental data [147] that the shape of a SIA plots (with close upper and lower limits) agree with the shape of a CRTA recorded at a similar constant reaction rate if the concerned reactions are fitted either by "*n* order" or diffusion kinetic models but not if an Avrami-Erofeev kinetic model is obeyed. It was proved [144] that in this latter case, the SIA control would force the reaction to take place at a constant temperature almost all over the α range. This behaviour can be understood bearing in mind

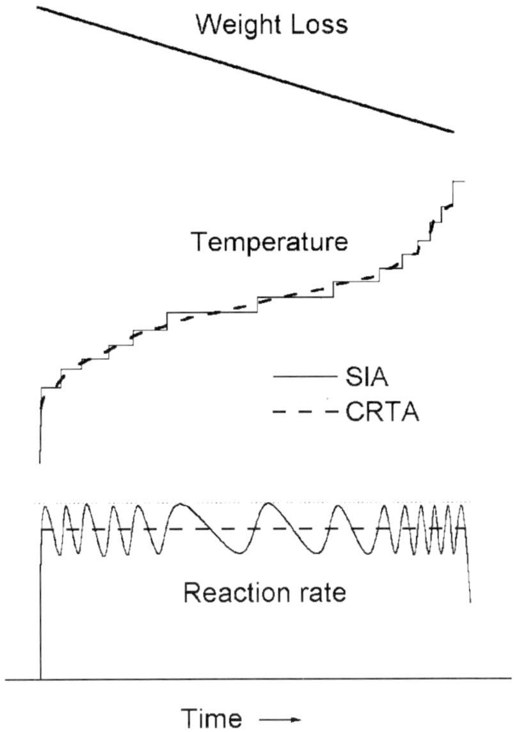

Figure 4-15. Comparison of SIA and CRTA curves.

that once the preset reaction rate is reached and the acceleratory period starts, the temperature will drop in order to compensate the abrupt increase of the reaction rate. CRTA permits to control the temperature in such a way that, if required, the α-T curve back on itself while SIA cannot and, as a consequence, the temperature remains constant while the measured reaction rate is higher than the upper preset limit, which practically occurs almost all over the α range. These theoretical conclusions were checked [147] by comparing the CRTA and SIA curves recorded under similar experimental conditions for the thermal decomposition of anhydrous nickel nitrate which is known to fit an A2 kinetic model. Figure 4-16 illustrates that the CRTA curve falls back on itself upon achieving the preset constant rate while the SIA curve becomes an isotherm at the temperature T attained at the moment at which the preset upper limit was reached. The comparison of the SIA and CRTA curves obtained by Arii and Fuji [126] for the thermal dehydration of calcium sulphate dihydrate, also shown in Figure 4-16, supports the above conclusion.

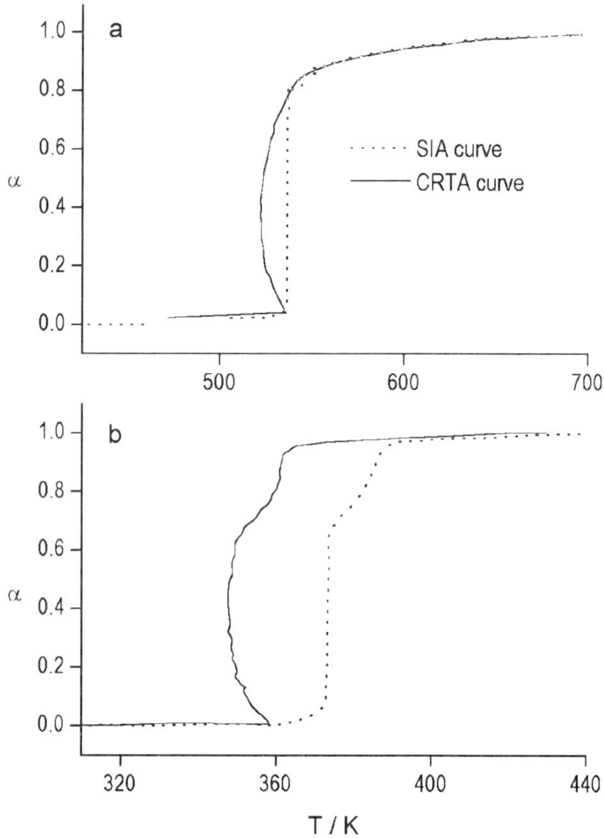

Figure 4-16. SIA and CRTA curves recorded for a) the decomposition of anhydrous nickel nitrate, and b) dehydration of calcium sulfate dihydrate.

4.10. A Unified Theory for the Kinetic Analysis of Solid State Reactions under any Thermal Pathway

As it has been shown above, the methodologies available for performing the kinetic analysis of isothermal, rising temperature and SCTA were independently developed and, thus, different sets of master plots were proposed for analysing isothermal, rising temperature, or SCTA data. A common strategy for the combined analysis of experimental data obtained from different experimental methods was missing and, even in the scarce number of works [20,21] in which both isothermal, linear heating and SCTA methods were applied at the time to the kinetic analysis of a given solid state reaction, the kinetic parameters were independently calculated and later compared among themselves for deciding the correct reaction mechanism. A unified theory that permits the combined analysis of the experimental data obtained under any thermal pathway has been recently outlined [22] and a set of generalized master plots were proposed departing from the *reduced time* Ozawa's concept [97].

The Ozawa reduced time, θ, is defined as

$$\theta = \int_0^t \exp\left(-\frac{E_a}{RT}\right) dt \tag{25}$$

that after differentiation becomes

$$\frac{d\theta}{dt} = \exp\left(-\frac{E_a}{RT}\right) \tag{26}$$

where θ denotes the reaction time taken to attain a particular α at infinite temperature.

Combining Eqs. (4) and (26) , the following expression is obtained:

$$\frac{d\alpha}{d\theta} = Af(\alpha) \tag{27}$$

or

$$\frac{d\alpha}{d\theta} = \frac{d\alpha}{dt} \exp\left(\frac{E_a}{RT}\right) \tag{28}$$

where dα/dθ corresponds to the generalized reaction rate.

Using as reference point α = 0.5, the following equations are easily derived from Eqs. (27) and (26), respectively

$$\frac{d\alpha/d\theta}{(d\alpha/d\theta)_{\alpha=0.5}} = \frac{f(\alpha)}{f(0.5)} \tag{29}$$

and

$$\frac{d\alpha/d\theta}{(d\alpha/d\theta)_{0.5}} = \frac{d\alpha/dt}{(d\alpha/dt)_{0.5}} \frac{\exp(E_a/RT)}{\exp(E_a/RT_{0.5})} \tag{30}$$

where the left hand side of Eq. (30) was named "reduced generalized reaction rate" and the suffix 0.5 refers to the values of the considered parameters at α = 0.5.

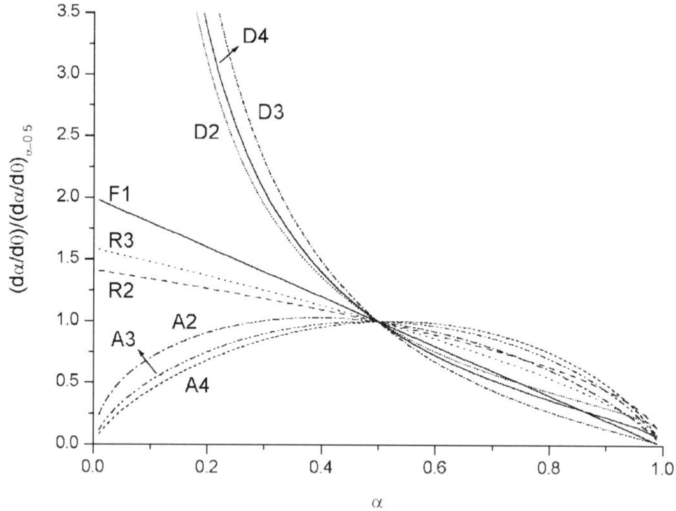

Figure 4-17. Generalized reduced rate master plots for different kinetic models.

The master plots of the "reduced generalized reaction rate" as a function of α for different kinetic models [22] are shown in Figure 4-17. These master plots would be indistinctly used for analysing experimental data obtained under any heating pathway (i.e., isothermal, rising temperature, SCTA, etc.), although the temperature conditions of the experimental kinetic data have to be taken into account. For the experimental kinetic data under isothermal conditions, both the exponential terms in Eq. (30) offset each other because in such a case $T = T_{0.5}$, so that the experimental master plot can be derived directly from a single isothermal curve of dα/dt against α.

On the other hand, for non isothermal data, the exponential terms in Eq. (30) cannot be cancelled out. For calculating the "reduced-generalized reaction rate" at a given α from non isothermal data under linear and non-linear heating, in addition to the kinetic data of a single measurement, the value of E_a should be previously known either from *Rate jump* or *Isoconversional methods*. As a special case of the nonlinear nonisothermal data, the ratio of rate terms in real time in Eq. (30) is to be unity for the kinetic data of CRTA. It should be remarked that for this particular case the master plots represented in Figure 4-18 are like those previously introduced in Figure 4-12, but those in Figure 4-12 were represented in logarithmic form.

The practical usefulness of these master plots was checked by using the experimental data of the isothermal mass-loss trace, rising temperature TG and CRTA obtained for the thermal decomposition of anhydrous $ZnCO_3$ (smithsonite) under high vacuum [20]. Figure 4-18 shows that independently of

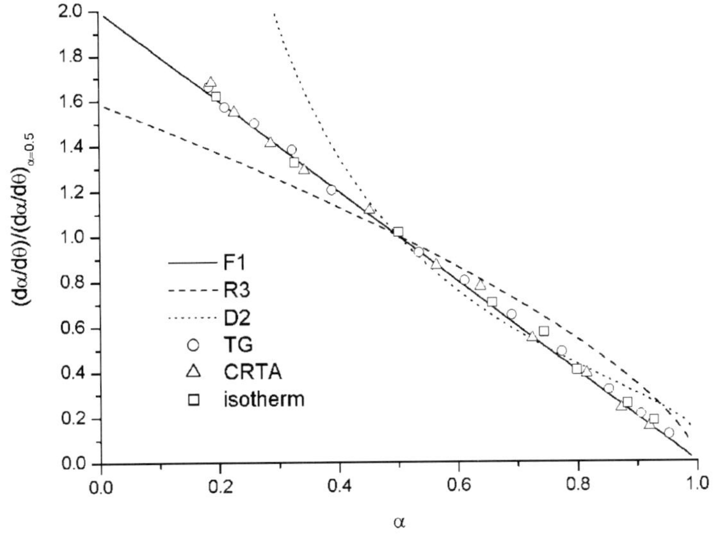

Figure 4-18. Comparison of the experimental generalized reduced rate obtained for the thermal decomposition of $ZnCO_3$ with theoretical master curves.

the heating pathway followed by the reaction, the experimental master plots are matching the theoretical master plots corresponding to the F1 kinetic model in agreement with the conclusion obtained from another source [20].

The above analysis suggest that it is possible to redefine the master plots to get universal plots which can be applied to the kinetic analysis of solid state reactions, whatever the thermal pathway used for recording the experimental data. Moreover, the results included in Figure 4-18 support the universal character of Eq. (4) for being used either under isothermal or non isothermal conditions in contradiction to the authors which question the applicability of Eq. (4) for the kinetic analysis of experimental data obtained under non isothermal conditions as discussed in section (2).

4.11. Comparison of the Resolution Power of CRTA and Conventional Non-Isothermal Methods: A Kinetic Approach

It has been claimed in the literature [112,79,80,138,139,148–153] that SCTA has a higher resolving power for discriminating mutually independent overlapping reactions than conventional non-isothermal methods. According to a recent review [37], "inspection of the available literature does seem to support

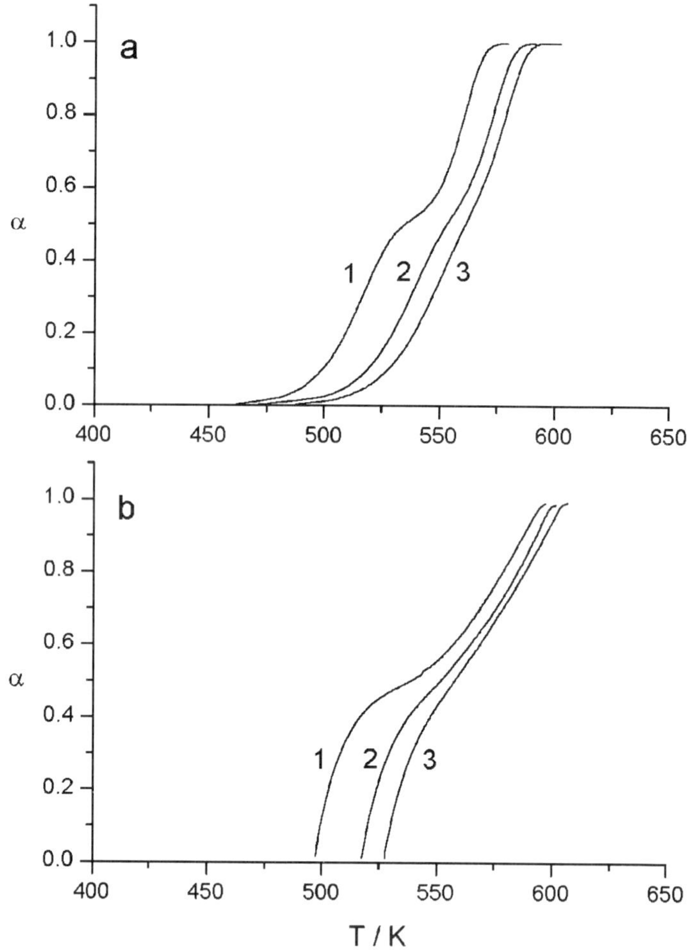

Figure 4-19. Simulated TG curves (a) at different heating rates: $\beta = 0.1$ (curve 1), $\beta = 0.5$ (curve 2), and $\beta = 1$ (curve 3); and simulated CRTA curves (b) at different constant reaction rates: $C = 7.69 \times 10^{-4}$ min^{-1} (curve 1), $C = 3.85 \times 10^{-3}$ min^{-1} (curve 2), and $C = 8 \times 10^{-3}$ min^{-1} (curve 3). The curves have been simulated for two overlapping first order reactions with the following kinetic parameters: $E_{a1} = 167$ kJ ml^{-1}; $A_1 = 6 \times 10^{14}$ min^{-1}; $E_{a2} = 334$ kJ mol^{-1}, $A_2 = 1.8 \times 10^{29}$ min^{-1}.

the assertion that resolution is improved" by SCTA. This can be simply understood as a consequence of the high level of control provided by CRTA: the rate of the reaction is measured with great accuracy (since it is constant), the uncertainty about the sample temperature is lowered at will (by imposing a low rate of reaction and by therefore decreasing at will the temperature gradients) and the surrounding atmosphere is also well controlled, thanks to the reduction of pressure gradients. In other words, CRTA simply provides, relatively automatically, the experimental conditions for meaningful, reliable, kinetic experiments. This does not mean that CRTA is the only way for that: *provided the level of control of the temperature and pressure gradients and of the surrounding atmosphere is comparable,* any isothermal or constant heating rate experiment can, in principle, bring the same information. This was shown by Criado et al. [154] who proposed to compare the resolution power of CRTA and conventional rising temperature from experiments for which the total time required for carrying out the reaction were the same. Figure 4-19 shows the sets of curves that were simulated for two independent "first order" reactions by selecting the values of the heating rate, β, of the rising temperature curves and the constant reaction rate, C, of the CRTA plots in such a way that the time consumed from the starting to the end of the process was the same for both rising temperature and CRTA curves. The inspection of the plots shown in Figure 4-19 suggests that both methods have almost the same resolution power if (i) β and C are selected in such a way that the total reaction time is the same for both CRTA and conventional rising temperature experiments and (ii) as pointed out above, the levels of control of the experimental conditions are comparable (which, in general, is the delicate point in conventional thermal analysis experiments).

4.12. Conclusions

It can be concluded from this review that SCTA methods are the most reliable approach for obtaining meaningful kinetic parameters of solid state reactions, as stated in a previous reviews [37]. This behaviour stands on the two following points: 1) SCTA methods allows a better control of the heat and mass transfer phenomena than isothermal or rising temperature methods making easier to determine activation energies really representative of the forward chemical reaction and 2) a single SCTA curve gives much more kinetic information on the reaction under study than a single either isothermal or linear rising temperature experiments. It has been also shown that the combined use of the *Rate jump* method and the master curves discussed in this review help to determine whether the process to be studied follows a single kinetic model even if this model is not represented by any of the "ideal" kinetic equations included in Table (1). Finally, it has been demonstrated that the methods of kinetic analysis of solid state reactions can be unified in order to allow the simultaneous analysis of experimental data whatever would be the thermal pathway used for recording the experimental data, though keeping in mind that it is probably with CRTA that the highest level of control of the experimental conditions can be automatically achieved.

References

1. J.P. McCallum and J. Tanner, Nature, 225 (1970) 1127.
2. A.L. Drapper, Thermochim. Acta, 1 (1970) 3(5).
3. J.R. McCallum, Nature, 232 (1971) 41.
4. V.M. Gorbatchev and V.A. Logvinenko, J. Thermal Anal., 4 (1972) 475.
5. P.D. Garn, J. Thermal Anal., 6 (1976) 237.
6. J. Norwiz, Thermochim. Acta, 25 (1978) 123.
7. A. Dutta and M.E. Ryan, Thermochim. Acta, 33 (1979) 387.
8. P.G. Boswell, J. Thermal Anal., 18 (1980) 353.
9. J. Blazejowsky, Thermochim. Acta, 48 (1981) 109.
10. H.Y. Yinnon and D.R. Uhlmann, J. Noncrystaline. Solids, 54 (1983) 253.
11. J.R. McCallum, Thermochim. Acta, 53 (1982) 375.
12. E. Koch and B. Stilkerieg, Thermochim. Acta, 33 (1979) 387.
13. Z. Chvoj, J. Sesták and A. Tríska, "Kinetic phase diagrams: Non-equilibrium phase transitions", Elsevier, Amsterdam 1991.
14. D. Dollimore and K.H. Tonge in "Reactivity of Solids", Proc. 5th Int. Symp., Elsevier, Amsterdsm 1965, p. 507.
15. F. Skvara and V. Satava, J. Thermal Anal., 2 (1970) 325.
16. D.W. Johnson and P.K. Gallagher, J. Phys. Chem., 76 (1972) 1474.
17. J.M. Criado, F. González and J. Morales, Thermochim. Acta, 12 (1975) 337.
18. J.M. Criado, A. Ortega, C. Real and E. Torres, Clay Min., 19 (1984) 653.
19. J.M. Trillo, G. Munuera and J.M. Criado, Catal. Rev., 7 (1972) 51.
20. F.J. Gotor, M. Macías, A. Ortega and J.M. Criado, Int. J. Chem. Kinetics, 30, (1998) 647.
21. F.J. Gotor, M. Macías, A. Ortega and J.M. Criado, Phys. Chem. Minerals, 27 (2000) 495.
22. F.J. Gotor, J.M. Criado, J. Málek and N. Koga, J. Phys. Chem. A, 104 (2001) 10777.
23. L.A. Pérez-Maqueda, J.M. Criado, F.J. Gotor and J. Málek, J. Phys. Chem. A, 106 (2002) 2862.
24. W.E. Brown, D. Dollimore and A.K. Galwey, "Reactions in the Solid State", Comprehensive Chemical Kinetics, Vol. 22, Elsevier, Amsterdam 1980.
25. P.D. Garn, Crit. Rev. Anal. Chem., 3 (1972) 65.
26. A.K. Galwey and M.E. Brown, "Kinetic background to thermal analysis and calorimetry", Handbook of Thermal Analysis and Calorimetry", Elsevier, Amsterdam 1998, Vol. 1, p. 147–224.
27. A.K. Galwey and M.E. Brown, "Thermal Decomposition of Ionic Solids", Elsevier, Amsterdam 1999.
28. E.V. Boldyreva, Thermochim. Acta, 110 (1987) 107.
29. J. Sesták and J. Kratochvil, J. Thermal Anal., 5 (1973) 193.
30. J. Sesták and J. Kratochvil, Thermochim. Acta, 7 (1973) 330.
31. J.V. Sesták, V. Satava and W.W. Wendlandt, Thermichim. Acta, 7 (1973) 333.
32. J.M. Criado, Thermochim. Acta, 43 (1981) 111.

33. J. Sestàk, Thermochim. Acta, 83 (1985) 391.
34. E. Segal, Thermochim. Acta, 148 (1989) 127.
35. J. Sesták, "Thermophysical Properties of Solids" Elsevier, Amsterdam, 1984.
36. M. Reading, "Controlled Rate Thermal Analysis and Beyond" in "Thermal Analysis – Techniques and Applications" (E.L. Charsley and S.B Warrington Eds.), Royal Society of Chemistry, Cambridge 1992, p.126–155.
37. M. Reading, "Controlled Rate Thermal Analysis and Related Techniques" in "Handbook of Thermal Analysis and Calorimetry" (P.K. Gallagher, General Ed.), Vol. 1: "Principles and Practice" (M.E. Brown, Ed. of Vol. 1), Elsevier, Amsterdam 1998, Vol. 1, p. 423–443.
38. J.M. Criado, F. Rouquerol and J. Rouquerol, Thermochim. Acta, 38 (1980) 109 and 117.
39. Y. Laureiro, A. Jerez, F. Rouquerol and J. Rouquerol, Thermochim. Acta, 278 (1996) 165.
40. J.M. Criado, Mater. Sci. Monogr., 6 (1980) 1096.
41. J.M. Criado, Thermal Analysis B. Miller (Ed.), Proc. 7th Int. Conf. Thermal Anal., Wiley, London 1982, Vol. 1, p. 99.
42. M. Reading, D. Dollimore, J. Rouquerol and F. Rouquerol, J. Thermal Anal., 37 (1984) 775.
43. J.M. Criado, A. Ortega, J. Rouquerol, and F. Rouquerol, Bol. Soc. Españ. Ceram. Vidr. 25, (1987) 407 (in Spanish).
44. A. Ortega, S. Akahouari, F. Rouquerol and J. Rouquerol, Thermochim. Acta, 163 (1990) 25.
45. J.M. Criado, M.J. Diánez, M. Macías y M.C. Paradas, Thermochim. Acta, 171 (1990) 229.
46. J.M. Criado and A. Ortega, J. Thermal Anal., 37, (1991) 2369.
47. M. Reading, D. Dollimore and R. Whitehead, J. Thermal Anal., 37 (1991) 2165.
48. H. Tanaka, Netsu Sokutey, 19 (1992) 32.
49. J. Málek, J. Sesták, F. Rouquerol, J. Rouquerol, J.M. Criado and A. Ortega, J. Thermal Anal., 38 (1992) 71.
50. A. Ortega, S. Akahouari, F. Rouquerol and J. Rouquerol, Thermochim. Acta, 235 (1994) 1535.
51. N. Koga, J.M. Criado, Int. J. Chem. Kinet., 30 (1998) 737.
52. T. Hatakeyama and L. Zhenhai, Handbook of Thermal Analysis, Wiley, Chichester 1998.
53. A. Finaru, I. Salageanu and E. Segal, J. Thermal Anal. Calorim., 61 (2000) 239.
54. J.M. Criado, A. Ortega, J. Rouquerol and F. Rouquerol, Thermochim. Acta, 240 (1994) 247.
55. J.M. Criado, M. González, J. Málek and A. Ortega, Thermochim. Acta, 254 (1995).
56. N. Koga and H. Tanaka, Thermochim. Acta, 388 (2002) 41.
57. F. Rouquerol, Y. Laureiro and J. Rouquerol, Solids State Ionics, 63–65 (1993) 363.

58. S. Bordère, F. Rouquerol, P.L. Llewellyn and J. Rouquerol, Thermochim. Acta, 282–283 (1996) 1.
59. J. Rouquerol, Bull. Soc. Chim. Fr., (1964) 31.
60. J. Rouquerol, J. Thermal Anal., 2 (1970) 123.
61. L.A. Pérez-Maqueda, J.M. Criado and F. Gotor, Intern. J. Chem. Kinetics, 34 (2002) 184.
62. J. Rouquerol, J. Thermal Anal., 5 (1973) 203.
63. J.M. Criado, Thermochim. Acta, 28 (1979) 307.
64. J.M. Criado and J. Morales, Thermochim. Acta, 16 (1976) 382.
65. J.M. Criado and J. Morales, Thermochim. Acta, 41 (1980) 125.
66. J.M. Criado, D. Dollimore and G. R. Heal, Thermochim. Acta, 54 (1982) 159.
67. J.H. Flynn, J. Thermal Anal., 34 (1988) 367.
68. R.K. Agrawal, Thermochim. Acta, 128 (1988) 185.
69. S.V. Vyazovkin and A.I. Lesnikovich, J. Thermal Anal., 35 (1989) 2169.
70. J.M. Criado, A. Ortega and F. Gotor, Thermochim. Acta, 157 (1990) 171.
71. N. Koga, J. Sesták and J. Málek, Thermochim. Acta, 188 (1991) 333.
72. J. Málek, Thermochim. Acta, 200 (1992) 257.
73. S. Vyazovkin and C.A. Wight, Int. Rev. Phys. Chem., 17 (1998) 407.
74. M.E. Brown et al. (ICTAC project), Thermochim. Acta, 355 (2000) 125.
75. J.M. Criado, J. Thermal Anal., 21, 155 (1981).
76. W.E. Garner, "Chemistry of the Solid State", Butterworths, London 1955.
77. M.J. Tiernan, P.A. Barnes and G.M.B. Parkes, J. Phys. Chem. B 2001, 105 (2001) 220.
78. J.M. Criado, A. Ortega and C. Real, React. Solids, 4 (1987) 93.
79. J. Paulik and F. Paulik, "Simultaneous thermoanalytical examinations by means of derivatograph", Wilson-Wilson's Comprehensive Analytical Chemistry, Elsevier 1981, Vol. XII.
80. F. Paulik, "Special Trends in Thermal Analysis", John Wiley, New York 1995.
81. P.A. Barnes, G.M.B. Parkes, D.R. Brown and E.L. Charsley, Thermochim. Acta, 269/270, (1995) 665.
82. F. Gomez, P. Vast, Ph Lewellyn and F. Rouquerol, J. NonCryst. Solids, 222 (1997) 415.
83. F. Gomez, P. Vast, P. Lewellyn and F. Rouquerol, J. Thermal Anal., 49 (1997) 1171.
84. E. Badens, P. Llewellyn, J.M. Fulconis, C. Jourdan, S. Veesler, R. Boistelle and F. Rouquerol, J. Solid State Chem., 139 (1998) 37.
85. F. Gomez, P. Vast, F. Baebieux, P. Lewllyn and F. Rouqueril, High Temperatures – High Pressures, 30 (1998) 575.
86. J.M. Fulconis, F. Morato, F. Rouquerol, E. Fourcade, A. Feugier and J. Rouquerol, J. Thermal Anal. Calorim., 56 (1999) 1443.
87. S. Ichihara, A. Endo and T. Arii, Thermochim. Acta, 360 (2000) 179.
88. N. Koga, J.M. Criado and H. Tanaka, J. Thermal Anal. Calorim., 60 (2000) 943.

89. E.A. Fesenko, P.A. Barnes, G.M.B. Parkes, E.A. Dawson and M.J. Tiernan, Topics in Catalysis, 19 (2002) 283.
90. E.A. Giess, J. Amer. Ceram. Soc., 46 (1963) 364.
91. J.H. Sharp, G.W. Brindley and B.N.N. Achar, J. Am. Ceram. Soc., 49 (1966) 379.
92. B. Delmon, "Introduction a la Cinétique Héterogéne" Publocations de Institute Francais du Pétrole, Paris 1969.
93. C.J. Keatch and D. Dollimore, "An Introduction to Thermogravimetry", Heyden, London 1969.
94. N. Koga and J.M. Criado, J. Am. Ceram. Soc., 81 (1998) 2901.
95. J. Málek, Thermochim. Acta, 355 (2000) 239.
96. V. Satava and F. Skavára, J. Am. Ceram. Soc., 52 (1966) 591.
97. T. Ozawa, J. Thermal Anal., 2 (1970) 325.
98. L. Jones, D. Dollimore and T. Nicklin, Thermochim. Acta, 1 (1975) 240.
99. J.M. Criado, Thermochim. Acta, 24 (1978) 186.
100. J.M. Criado, J. Málek and A. Ortega, Thermochim. Acta, 147 (1989) 377.
101. J. Málek, Thermochim. Acta, 200 (1992) 257.
102. J. Málek and J.M. Criado, Thermochim. Acta, 236 (1994) 187.
103. J. Málek, Thermochim. Acta, 267 (1995) 61.
104. N. Koga and J. Sesták, J. Am. Ceram. Soc., 83 (2000) 1753.
105. J. Málek, T. Mitsuhashi and J.M. Criado, J. Mater. Res., 16 (2001) 1862.
106. M. Reading, Thermochim. Acta, 135 (1988) 37.
107. L.A. Pérez-Maqueda, A. Ortega and J.M. Criado, Thermochim. Acta, 277 (1996) 165.
108. J.M. Criado, M. González, A. Ortega and C. Real, J. Thermal Anal., 29 (1984) 93.
109. F. Rouquerol and J. Rouquerol, in "Thermal Analysis", Proc. 3rd ICTA, 1971, Davos (H.G. Wiedemann, Ed.), Birkhäuser Verlag, Basel 1972, Vol. 1, p. 373.
110. F. Rouquerol, S. Régnier and J. Rouquerol, in "Thermal Analysis", Proc. 4th ICTA, 1974, Budapest (I. Buzas, Ed.), Akademiai Kiado and Heyden and Sons, Budapest 1975, Vol. 1, p. 313.
111. J. Rouquerol, Pure Appl. Chem., 57 (1985) 69.
112. S. Bordère, F. Rouquerol, J. Rouquerol, J. Rouquerol, J. Estienne and A. Floreancing, J. Thermal Anal., 36 (1990) 1651.
113. J. Rouquerol, S. Bordère and F. Rouquerol, in "Thermal Analysis in the Geosciences", (W. Smykatz-Kloss and S.J. Warne, Eds.), Springer Verlag, Berlin 1991, p. 134.
114. J.M. Criado, in "Thermal Analysis", Proc. 6th ICTA, Bayreuth, 1980 (H.G. Wiedemann, Ed.), Birkhauser Verlag, Basel 1980, p. 1(5).
115. P.A. Barnes, G.M.B. Parkes and E.L. Charsley, Anal. Chem., 66 (1994) 2226.
116. Y. Laureiro, A. Jerez, F. Rouquerol and J. Rouquerol, Thermochim. Acta, 278 (1996) 165.
117. P. Dion, J.F. Alcover, F. Bergaya, A. Ortega, P.L. Llewellyn and F. Rouquerol, Clay Miner., 33 (1998) 269.
118. G.M.B. Parkes, P.A. Barnes and E.L. Charsley, Anal. Chem., 71 (1999) 2482.

119. M.J. Tiernan, P.A. Barnes and G.M.B. Parkes, J. Phys. Chem. B, 103 (1999) 6944.
120. M.J. Tiernan, P.A. Barnes and G.M.B. Parkes, J. Phys. Chem. B, 103 (1999) 338.
121. P.A. Barnes, M.J. Tiernan and G.M.B. Parkes, J. Thermal Anal. Calorim., 56, (1999) 733.
122. M.J. Tiernan, P.A. Barnes and G.M.B. Parkes, J. Phys. Chem. B, 105 (2001) 220.
123. E.A. Fesenko, P.A. Barnes, G.M.B. Parkes, D.R. Brown and M. Naderi, J. Phys. Chem. B, 105 (2001) 6178.
124. M.J. Tiernan, E.A. Fesenko, P.A. Barnes, G.M.B. Parkes and M. Ronane, Thermochim. Acta, 379 (2001) 163.
125. M. Reading, J. Thermal Anal. Calorim., 64 (2001) 7.
126. T. Arii and N. Fujii, J. Anal. Appl. Pyrolysis, 39 (1966) 129.
127. X.E. Cai, H. Shen, C.H. Zhang, Y.X. Wang and Z. Kong, J. Thermal Anal. Calorim., 60 (2000) 623.
128. N. Koga, J.M. Criado and H. Tanaka, Netsu Sokutey, 27 (2000) 128.
129. A. Ortega, Thermochim. Acta, 298 (1997) 205.
130. J. Rouquerol, Thermochim. Acta, 144 (1989) 209.
131. M.E. Brown, J. Thermal Anal. Calorim., 49 (1997) 17.
132. J. Málek, L. Tichý and J. Klokorka, J. Thermal Anal., 33 (1988) 667.
133. J. Málek, Thermochim. Acta, 129 (1988) 293.
134. J. Málek, J. Nin-Crystalline Solids, 107 (1989) 323.
135. J. Málek, J. Thermal Anal., 40 (1993) 159.
136. J. Málek, Y. Messaddeq, S. Inoue and T. Mitssuhashi, J. Mater. Sci., 30 (1995) 3082.
137. J.M. Criado, L.A. Pérez-Maqueda and A. Ortega, J. Thermal Anal., 41 (1994) 1535.
138. J. Paulik and F. Paulik, Anal. Chim. Acta, 56 (1971) 328.
139. F. Paulik and J. Paulik, Thermochim. Acta, 100 (1986) 23.
140. O.T. Sorensen, Thermochim. Acta, 50 (1981) 163.
141. O.T. Sorensen, J. Thermal Anal. Calorim., 56 (1999) 17.
142. P.S. Gill, S.R. Sauerbrunn and B.S. Crowe, J. Thermal Anal., 38 (1992) 255.
143. I.M. Salin and J.C. Seferis, J. Appl. Polymer. Sci., 47 (1993) 847.
144. V. Berbeni, A. Marini and G. Bruni, Thermochim. Acta, 322 (1998) 137.
145. C.P. Jaroniec, M. Kruk, M. Jaroniec and A. Sayari, J. Phys. Chem., 102 (1998) 5503.
146. A. Zanier, J. Thermal Anal. Calorim., 64 (2001) 377.
147. F.J. Gotor, L.A. Pérez-Maqueda, A. Ortega and J.M. Criado, J. Thermal Anal., 53 (1998) 389.
148. M. Ginicmarkovic, N.R. Choudhury, M. Dimopulos, D.R.G. Williams and J. Matisons, Thermochim. Acta, 316 (1998) 87.
149. R.G. Ford and P.M. Bertsch, Clays Clay Miner., 47 (1999) 329.
150. J. Choma, W. Burakiewiczmortka, M. Jaroniek, Z. Li and J. Klinik, J. Colloid Interface Sci., 214 (1999) 438.

151. F. Paulik, J. Thermal Anal. Calorim., 58 (1999) 711.
152. F. Paulik, Thermochim. Acta, 341 (1999) 711.
153. Z. Ding and R.L. Frost, Thermochim. Acta, 389 (2002) 185.
154. J.M. Criado, A. Ortega and F. Gotor, Thermochim. Acta, 203 (1992) 18.

Chapter 5

SCTA AND CERAMICS

O. TOFT SORENSEN[1]

with contribution from

J.M. CRIADO[2]

1- Risoe National Laboratory, Roskilde, Denmark
2-Instituto de Ciencia de Materiales, Centro Coordinado C.S.I.C-Universidad de Sevilla, Sevilla, Spain

5.1. Introduction

Conventional Thermal Analysis has since its introduction been used abundantly in studies of classical ceramics (clay products, porcelain, concrete) to establish the optimum process conditions for the fabrication of these materials and to measure their properties and behaviour. As a relatively new technique, however, the main applications of Sample-Controlled techniques (SCTA) has been in more fundamental studies of what generally is termed as engineering ceramics (structural and functional ceramics) and in this chapter we shall therefore focus on some recent thermogravimetric and dilatometric studies on these materials. Although many catalyst and adsorbents also can be considered as engineering ceramics, SCTA studies on these materials are however treated separately in Chapters 6 and 7.

5.2. Sample Controlled Thermogravimetry

5.2.1. STEPWISE ISOTHERMAL ANALYSIS OF BA-OXALATE [1]

Thermal Decomposition

Thermal decomposition studies of oxalates are of importance in synthesizing many double- or multicomponent oxides. High purity barium titanates, barium cerates and barium zirconates, which all are important electroceramic materials, can for instance be prepared from mixed metal oxalate compounds [2–5]. Accordingly, it is of great interest to investigate the thermal decomposition of barium oxalate hemihydrate.

Figure 5-1 show the TG-DTG signals obtained in a conventional non-isothermal TA measurement on barium oxalate hemihydrate ($BaC_2O_4 \cdot 0{,}5\ H_2O$) by using a heating rate of 2°C min^{-1}[1].

The TG-DTG curves obtained using other heating rates showed similar shapes as shown in Figure 5-1 indicating that the thermal decomposition generally takes place in the following five steps:

$$BaC_2O_4 \cdot 0{,}5\ H_2O \cdot BaC_2O_4 + 0{,}5\ H_2O \quad (1)$$

$$BaC_2O_4 \cdot BaCO_3 + CO \quad (2)$$

$$BaCO_3 \cdot BaO + CO_2 \quad (3,4,5)^*$$

*detailed reactions involved given below

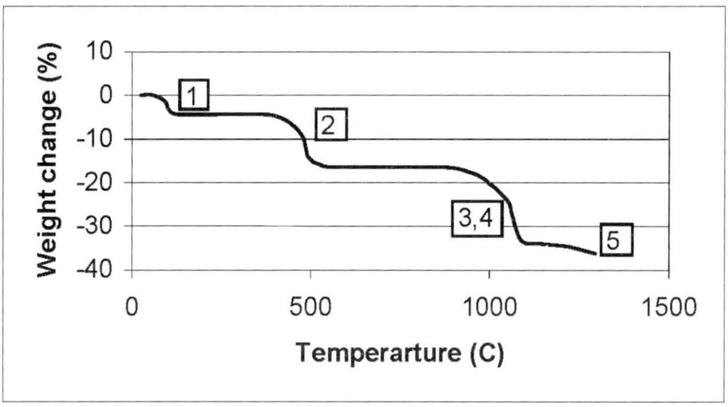

Figure 5-1. TG-DTG diagrams of BaC_2O_4 $0{,}5H_2O$ in air studied by non-isothermal thermogravimetry with a heating rate of 2°C min^{-1}. The numbers indicate the reactions 1-5.

The TG-DTA signals obtained using the SIA technique is shown in Figure 5-2 and again five distinct mass loss steps was obtained but in this case the decompositions took place isothermally and a much better resolution was obtained for the final decomposition of $BaCO_3$ in steps 3, 4 and 5.

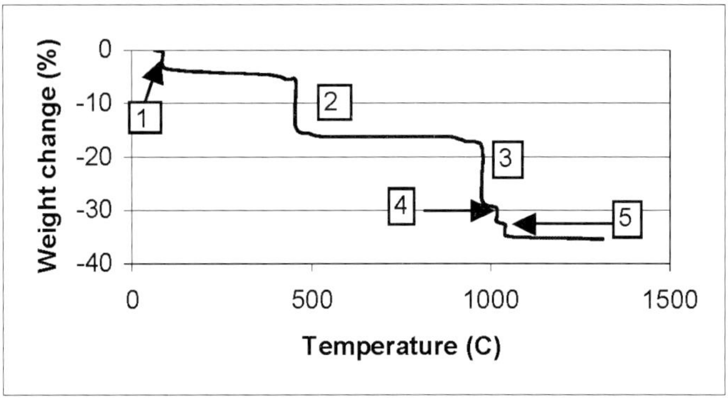

Figure 5-2. TG-DTA diagrams of BaC_2O_4 0,5 H_2O in air studied by stepwise isothermal analysis. The numbers indicate reactions 1-5.

From this curve it is thus possible to perform quantitative calculations on all steps and based on these calculations the following subdivision of the last three mass change steps were proposed:

$$3BaCO_3 \cdot BaCO_3(BaO)_2 + 2CO_2 \tag{3}$$

$$BaCO_3(BaO)_2 \cdot (BaCO_3)_{0,5}(BaO)_{2,5} + 1/2CO_2 \tag{4}$$

$$(BaCO_3)_{0,5}(BaO)_{2,5} \cdot 3BaO + 1/2\ CO_2 \tag{5}$$

This work thus indicates that two intermediate compounds with the compositions $BaCO_3(BaO)_2$ and $(BaCO_3)_{0,5}(BaO)_{2,5}$ are formed during the decomposition of $BaCO_3$. The thermal decomposition of $BaCO_3$ has also been studied by Kelly and Anderson [6,7] but with the aim of determining some of the thermodynamic properties of BaO and $BaCO_3$. Following Finkelstein [8] they assumed the existence of a basic carbonate, $BaCO_3BaO$, in the solid phase since they observed an eutectic in the $BaCO_3$-BaO system. Later on, however, Hackspill and Wolf [9] and Baker [10] assigned the composition $2BaCO_3BaO$ to this eutectic. The existence of these compounds.were however not confirmed in the present work.

Kinetic Studies

Non-isothermal thermogravimetry with a linear heating rate has been widely used as a tool for deriving kinetic parameters such as the activation energy and the pre-exponential factor in the Arrhenius equation. The advantages of this technique are obviously that the kinetics can be etstablished over the entire temperature range in a continous manner and that it is less time-consuming, but as shown above and in the previous sections, it has the great disadvantage that it cannot separate close-lying reactions. Combining SIA and non-isothermal techniques, however, is a good tool for precise kinetic studies as the reaction mechanism can be determined accurately from the isothermal data whereas the kinetic parameters can be determined from the non-isothermal data once the mechanism is known. This was demonstrated by Chen and Sorensen [1] in their kinetic study of the thermal decomposition of barium oxalate hemihydrate.

Isothermal Kinetics: As shown in Chapter 4, the kinetic equation describing a reaction under isothermal conditions can be expressed in the integral form as:

$$\kappa t = g(\alpha) \tag{6}$$

or the differential form as:

$$\mathrm{d}\alpha/\mathrm{d}t = \kappa f(\alpha) \tag{7}$$

where $g(\alpha)$ and $f(\alpha)$ are the integral and differential forms of α, the fraction decomposed in tme t, and κ is the rate constant as expressed by the Arrhenius equation:

$$\kappa = A \exp\{-E_a/\mathrm{RT}\} \tag{8}$$

where A is the pre-exponential factor (frequency factor) and E_a is the activation energy.

The form of $g(\alpha)$ and $f(\alpha)$ depend on the type of rate-limiting process, i.e. the slowest process, that controls the overall reaction. As it is clearly seen from Eq. (6) and (7) it is possible to evaluate the controlling mechanism from the experimental data either by plotting g(α) vs. t or f(α) vs. dα/dt, which should give a straight line for the correct mechanism. The equations used in the kinetic study on barium oxalate hemihydrate [11] are presented in Table 1 by Chen et al. in [1].

The data from the SIA measurements for the reactions (1)–(5) were fitted by the linear least-squares method using the different forms of g(α) and f(α) listed in the table by Chen et al. The controlling mechanism was determined by the best fitting straight line with the highest correlation coefficient (r) from the g(α) vs. t and the f(α) vs dα/dt plots. It is generally possible to visually choose the best straight line, but the coorrelation coefficient, r, gives a quantitative value of judgement. The values obtained for the different reaction mechanisms are are summarized in Table 3 again by Chen et al. for the five reactions both for the integral and the differential form of the kinetic equations.

It is obvious that for the proper functional forms, the results obtained by the two forms should closely agree. From this table previously published [1], this condition is best satisfied when the equation – $f(\alpha) = (1-\alpha)^{1/2}$, $g(\alpha) = 2[1-(1-\alpha)^{1/2}]$ – is used for all five isothermal steps. From this study it can therefore be concluded that the thermal decomposition of barium oxalate hemihydrtae, which takes place in five steps, most probably is a phase-boundary controlled process which takes place in two dimensions (cylindrical symmetry).

Non-isothermal Kinetics: It has been pointed out [12] that both the integral and the differential form can be used for non-isothermal cases. Since the measurement generally is performed at constant heating rate, $q = dT/dt$, substituion of $dt = dT/q$ and the rate constant κ into the differential equation (Eq. (7)) gives:

$$d\alpha/dT = (A/q)\exp[-E/RT]\, f(\alpha) \tag{9}$$

which by rearrangement and by taking the logaritm becomes:

$$\ln[(d\alpha/dT)\,(1/f(\alpha))] = \ln(A/q) - E/RT \tag{10}$$

This is termed the differential equation since it employs the differential form $f(\alpha)$. By plotting $\ln[(d\alpha/dT)(1/f(\alpha)]$ vs. $1/T$ one will obtain a straight line when the correct mechanism is chosen. The activationn energy E and the pre-exponential factor A, respectively, can then be calculated from the slope and the intercept of this line.

By integrating Eq. (9) between proper limits the Coats and Redfern equation [13], which is termed as the integral equation, can be derived:

$$\ln\,[g(\alpha)/T^2] = \ln(AR/qE)[1 - 2RT/E] - E/RT \tag{11}$$

and plotting $\ln[g(\alpha)/T^2]$ vs. $1/T$ a straight line will be obtained for the correct mechanism. The slope of this line equals $-E/RT$ and the intercept $\ln[(AR/qE)(1 - 2RT/E)]$, and when E is known A can also be calculated. In this study the average temperature for the reaction rate was used for the calculation of the pre-exponential factor A.

As an example, analysis of the kinetic data is carried out on the steps involved in the dehydration of barium oxalate hemihydrate. Using the possible forms of $f(\alpha)$ and $g(\alpha)$ the data obtained were analyzed by means of Eqs. (10) and (11) and the results are presented in Table 4 published by Chen et al. [1] and from which it is clear that the values of E and ln A obtained by the two methods are very close to each other. Furthermore it will be noted that the r-values obtained also confirm that this reaction is two-dimensionally phase-boundary controlled.

Finally Table 5-1 show the results obtained in analysing all steps in this way. Again these results confirm that the most probable mechanism for all steps is two-dimensional phase-boundary movement – the R2 reaction (Table 1 in [1]) – as also concluded from the analysis of the SIA data. The activation energy for the dehydration process was determined to be 87.0 ± 0.7 $kJmol^{-1}$, which is close to that found for the dehydration of calcium oxalate monohydrate, which is 88.44 $kJmol^{-1}$ [14]. The activation energy for the overall breakdown of barium oxalate was found to be 296.01 $kJmol^{-1}$, which is also close to that of thermal decomposition of anhydrous calcium oxalate, 282.44 $kJmol^{-1}$ [15]. Finally the activation energy for the decomposition of $BaCO_3$ to BaO has previously by Judd and Pope [16] to be 283 $kJmol^{-1}$ whereas we in this work for the first time found that the $BaCO_3$ decomposes in three steps with increasing activation energies of 298.2 ± 14.3, 318.0 ± 11.3 and 332.1 ± 9.8 $kJmol^{-1}$, respectively. The activation energy thus increases for the three steps indicating an increasing difficulty in the decomposition of $BaCO_3$.

Table 5-1. Kinetic parameters for the decomposition of $BaC_2O_4 \cdot 0{,}5\ H_2O$ by the analysis of data from the non-isothermal thermogravimtry

Decomp. step	E (kJ/mol)			ln A (s^{-1})			Mechanism
	Differential	Integral	Average	Differential	Integral	Average	R2
1	86,27	87,66	87,0	20,73	20,40	20,6	R2
2	310,58	281,38	296,0	42,85	37,97	40,3	R2
3	283,84	312,53	298,2	19,08	23,60	21,3	R2
4	306,74	329,34	318,0	40,17	43,28	41,7	R2
5	341,90	322,32	332,1	20,08	22,41	21,2	R2

5.2.2. STEPWISE ISOTHERMAL ANALYSIS OF PURE AND DOPED CE-CARBONATES

Pure cerium oxide and several of the rare earth doped cerium oxides exhibit high oxygen ion conductivities and they are therefore interesting materials for the application as solid electrolytes in low temperature (<600°C) oxygen sensors and fuel cells [17]. Very promising candidates in the late eighties were gadolinia and europia doped cerium oxides, which at that time had the best oxygen ion conduction ever known.

An important process for the fabrication of powders of these oxides was at that time a homogeneous precipitation of carbonates by hydrolysis of urea in the appropriate nitrate solutions followed by a calcination [18]. As the sintering properties of the oxide powders strongly depend on the calcination conditions, a thorough SIA study was performed on the kinetics of the thermal decomposition both on pure and on doped carbonates of several compositions.

SIA on Pure Ce-carbonate [19]

As already shown in Figure 3-11 and in [19] the thermal decomposition of the pure Ce-carbonate ($Ce_2O(CO_3) \cdot H_2O$) generally takes place in either one or several steps depending on the atmosphere. The following reactions were proposed for these steps:

(1) In air-one step at relatively low temperature:

$$Ce_2O(CO_3)H_2O + 1/2\ O_2 \longrightarrow 2CeO_2 + 2CO_2 + H_2O \quad (12)$$

where Ce^{+3} is oxidized to Ce^{4+} by atmospheric oxygen.

(2) In the neutral atmosphere, He – many small steps at low temperature, followed by two distinct steps at higher temperature for which the following reactions are proposed:

$$Ce_2(CO_3)_2\ H_2O \longrightarrow Ce_2O_2(CO_3) + CO_2 + H_2O \quad (13)$$

i.e. direct decomposition into a peroxide followed by disproportionation of this peroxide into CeO_2 at higher temperature:

$$Ce_2O_2(CO_3) \longrightarrow 2CeO_2 + CO \quad (14)$$

(3) In CO_2 – same reactions as in He, but reaction temperatures for (13) and (14) are higher.

(4) In CO – same scheme as for CO_2 but the temperatures are slightly different.

(5) The kinetic study was performed using the integral form for the kinetic equation, Eq. (6), as for the study on the thermal decomposition of barium oxalate hemihydrate discussed in the previous section. Plotting the integral function $F11(\alpha) = g11(\alpha) = 1/2(-\ln(1-\alpha))2$ vs. time for the decomposition in air a nice straight line was obtained (Fig. 5-3) indicating that the controlling reaction for this reaction is nucleation followed by two-dimensional growth (cylindrical symmetry) controlled by phase boundary movement. In contrast the to mechanism controlling the decompostion of barium oxalate nucleation thus play a significant role in the decomposition of the cerium carbonates.

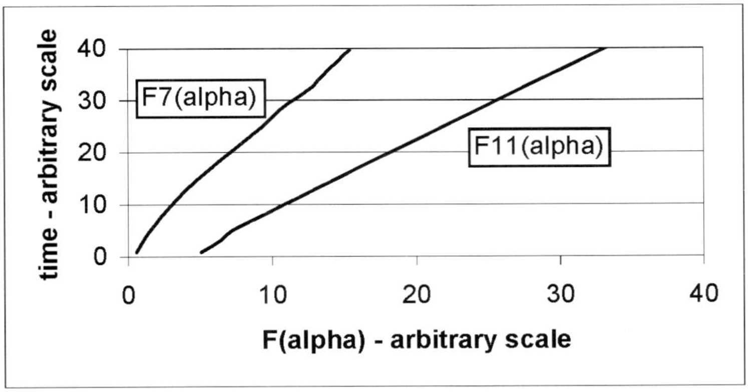

Figure 5-3. F(α) vs. time for isothermal de-composition of Ce-carbonate in air.

Another controlling mechanism was however observed when the thermal decompostion took place in CO. In this case the best straight line was observed using the $F7(\alpha) = 2(1 - (1-\alpha)^{1/2}$ equation (Fig. 5-4) indicating that the decomposition is a phase boundary movement controlled growth in this atmosphere.

SIA on Rare Earth Doped Cerium Carbonates

Figure 5-5 shows the SIA plots in air for Gd-doped Ce-carbonate at an intermidiate Gd content (for the curves for other compositions – see [18]) as well as the curves for pure Ce- and Gd-carbonate, whereas the SIA curve obtained for pure Eu-carbonate is shown in Figure 5-6 [18].

As will be noted, the precipitated $Ce_{1-x}RE_x$-carbonates ($0 < x < 1$) decomposed in one large well defined step followed by a slow continous weight loss as the temperature increased. Furthermore the magnitude of the first weight loss decreased with increasing x, whereas the temperature for this weight loss increased with x up to $x = 0.9$, after which it decreased. Assuming that the completely calcined powders had the composition $Ce_{1-x}RE_xO_{2-x/2}$ the composition calculated from the weight loss curves for the doped carbonates shown in these figures was $(Ce_{1-x}RE_x)_2O(CO_3)_2$ 1.0 H_2O corresponding to the stoichiometric formula for the cerium-carbonates.

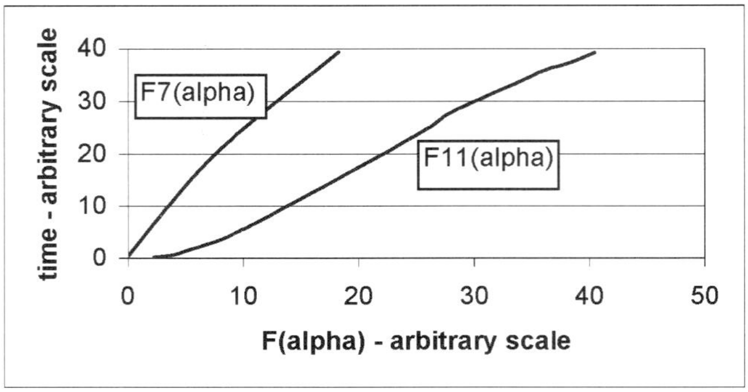

Figure 5-4. F(α) vs. time plots for curve (b) in CO-atmosphere. The best linear correlation is obtained with F7(α).

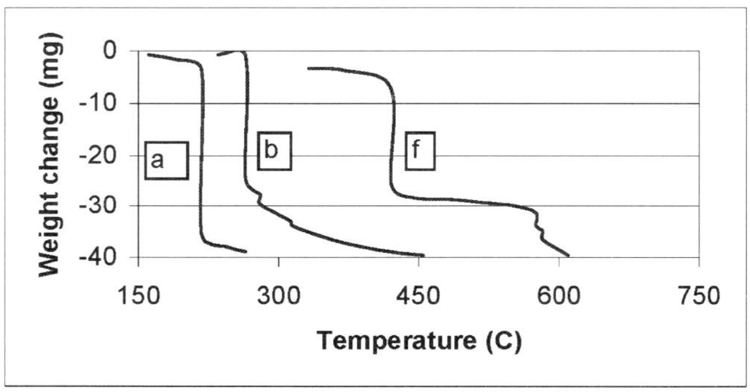

Figure 5-5. Weight loss versus temperature plots for some Ce-Gd-carbonates recorded by stepwise isothermal analysis. a: Ce-carbonate; b: $Ce_{0,7}Gd_{0,3}$ carbonate; f: Gd-carbonate.

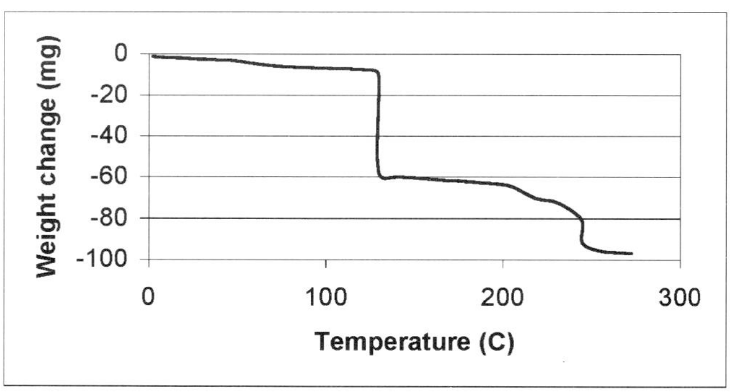

Figure 5-6. Weight loss versus temperature plot for Eu-carbonate recorded by step wise isothermal analysis.

As will be noted from the two figures, the first large weight loss for the Ce-, Eu-, and Gd-carbonates took place in one isothermal step in the SIA. Using the same method as discussed for the pure Ce-carbonates the rate determining steps for the Gd- and Eu-carbonates was found to follow the $F(\alpha) = 2(-\ln(1-\alpha))^{1/2}$ function which shows that the controlling reaction also in these carbonates are nucleation followed by two dimensional growth (cylindric symmetry) controlled by phase boundary movement. This mechanism was also found to control the decomposition both of the pure Gd- and Eu-carbonates.

5.2.3. FORCED STEPWISE ISOTHERMAL ANALYSIS: ACTIVATION ENERGIES FOR DECOMPOSITION OF CE-CARBONATES

A serious drawback with SIA is that it is not possible to determine the activation energies for a reaction from the isothermal data obtained by this technique. To overcome this problem the FSIA (Forced Stepwise Isothermal Analysis), was introduced [19]. A more detailed description of this technique has already been presented in Chapter 3 and in this section we shall therefore present and discuss the results obtained in a FSIA study using the thermal decomposition of Ce-carbonate as an example.

Figure 3-13 show the FSIA curves obtained for Ce-carbonate in various atmospheres. The characteristic feature of this technique is that the reaction is forced to take place in isothermal steps at increasing temperatures, which is clearly evident from the curves – the algorithm used forces the temperature to increase a few degrees after a weight change of about 20%.

In accordance with the kinetic study performed on the data obtained by SIA discussed in the previous section, straight lines were also obseved by plotting the two $F(\alpha)$ functions vs. time, but in this case in segments with characteristic slopes depending on the temperature as shown in Figure 5-7.

As the slope of the segments is equal to the Arrhenius constant k an Arrhenius plot, lnk vs. 1/T, can easily be constructed from these slopes. The resulting Arrhenius plots for the decomposition in air and CO respectively is shown in Figure 5-8 from which it will be noted that nice and straight lines were obtained in both cases.

Finally the activation energy and the pre- exponential factors can easily be calsulated from the slope and the intercepts of these lines and the results obtained in this way are presented in Table 5-2.

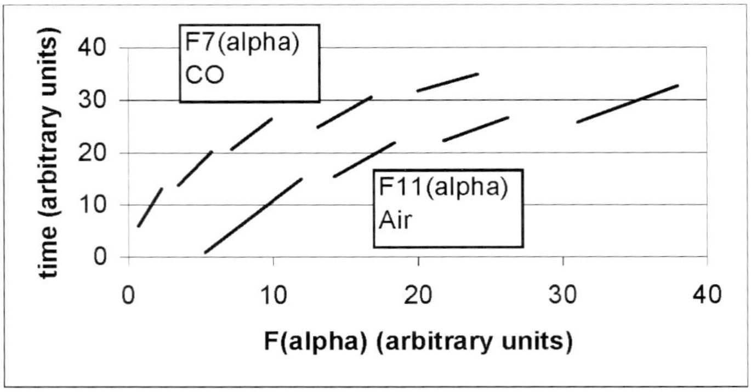

Figure 5-7. Arrhenius plot obtained from FSIA data shown in Figure 3-13.

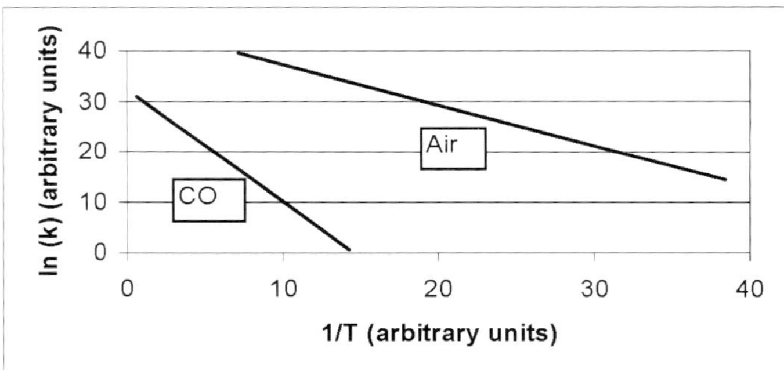

Figure 5-8. F(α) vs. time plot for FSIA data obtained in air and CO.

Table 5-2. Activation energies and pre-exponential factor for the decomposition of Ce-carbonate in air and CO

Atomosphere	E_a (kj/mole)	A (min^{-1})
Air (reaction a)	104.9	5.91×10^{9}
CO (reaction b)	297.8	5.03×10^{20}
CO (reaction c)	407.5	2.63×10^{26}

Finally it should be noted that the activation energies found by FSIA are quite consistent with the values determined for the thermal decomposition of other carbonates [11].

5.2.4. BINDER REMOVAL STUDIED BY SCTA

SIA

Generally ceramic parts are formed from ceramic powders using either dry pressing, different casting techniques such as slipcasting or injection moulding and plastic forming such as extrusion. In all cases it is necessary to add organic binders and lubricants to the powders in order to obtain dense and homogeneous specimens. In the case of injection moulding and extrusion large amounts of organic materials must be added and especially for these techniques it is very important that the organic materials are removed before the following high temperature sintering process, otherwise the specimens will severely crack during the sintering. Even the binder removal must be performed with great care by subjecting the parts to a controlled heating programme which depend on the composition of the binder mixture used.

The SIA technique has proved to be a powerful tool to establish the optimum conditionms for binder removal from injection moulded parts [20]. One example is shown in Figure 5-9, which gives the SIA-curve, i.e. the weight loss versus temperature, for the removal of a characteristic binder system. As will be noted, the decomposition temperatures of the compounds involved can be easily determined from this curve from which the optimum binder removal conditions also can be established. Compared to the binder burnout conditions used in conventional methods, the total burnout time can also be reduced considerably without crack formation using SIA.

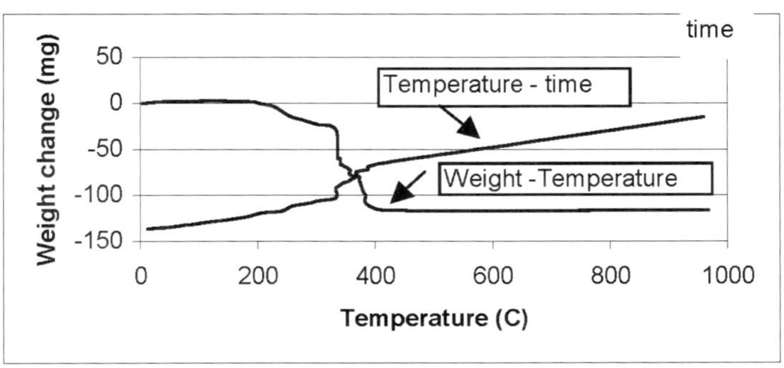

Figure 5-9. SIA curve for binder removal from injection moulded ceramic part.

The SIA technique has also been used industrially to control a binder removal furnace. The computer programme developed at Risoe National Laboratory, Denmark was used in this furnace system.

Binder Removal Studied by CRTA

In a recent study Arii et al. [39] showed that optimum conditions for binder burnout of ceramic green compacts could be determined with great confidence by using constant rate SCTA. In this work hard materials consisting primarily of 70–80% WC, 1–2% TaC and 10% Co was studied. The binder content was about 8% and the binder was composed of 1–2% ethyl alcohol, 4% plasticizer and 2–3% polyvinyl resin. The compacts were prepared by extrusion. The binder burnout measurements were performed in an inert helium atmosphere.

One advantage with this technique is that it enables a more precise control of the uniformity of the sample environment, i.e. the pressure of the gaseous decomposition products, the sample temperature and the pressure gradients within the sample. These conditions are therefore especially important to control in order to prevent the formation of defects in the samples during the debinding process. Comparing the microstructures of samples obtained using this technique with microstructures of samples treated in the conventioanl way, Arii et al. showed that flawless samples could indeed be obtained using the CRTA technique, which was not the case when burnout was performed using the conventional technique. Furthermore they also showed that the total burnout time is considerably reduced by this technique compared to the time involved in conventional debinding treatments.

5.3. Dilatometric SCTA Measurements

5.3.1. INTRODUCTION

The stepwise isothermal technique (SIA) has also proved to be useful in dilatometric studies of the sintering behaviour of ceramic compacts. By this technique, which generally is designated as Stepwise Isothermal Dilatometry (SID) – in early studies it was designated Quasi-isothermal dilatometry (QID) [25] – the temperature is controlled by the change in length of the sample – corresponding to the shrinkage due to the densification during the sintering – in the same way as for the thermogravimetric SIA technique. The sintering (densification) thus takes place in characteristic isothermal steps as observed for the thermal decompositions discussed in the previous sections. A typical SID-curve obtained for sintering of yttria doped zirconia [20] in the temperature range 850 to 1450°C is shown in Figure 5-10.

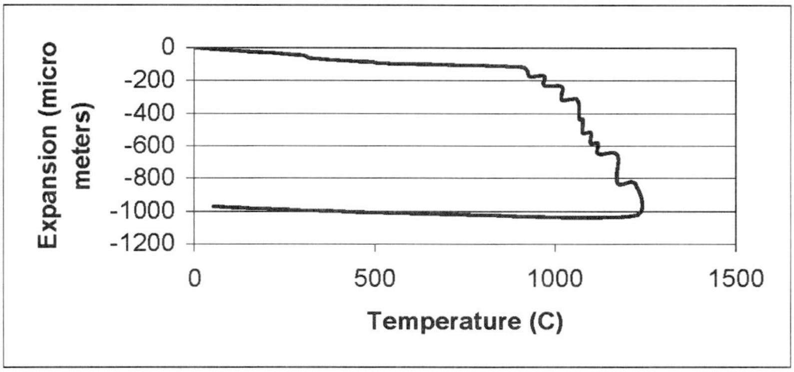

Figure 5-10. SID curve for zirconia ceramics (specimen TZ3YA-07).

The densification of a powder compact generally takes place at increasing temperatures in several successive and to some extent overlapping stages:

1) The first stage, where formation of necks between the particles is the main process. The densification during this stage is very small.
2) During the second stage, the growth of the necks continues and the irregular pores of the compact gradually changes into a circular form. The densification becomes more important during this stage.
3) In the third stage two processes are considered to compete: 1) pore elimination along grain boundaries formed during neckgrowth and 2) grain growth, which tend to isolate the pores inside the grains and which therefore slow down the pore elimination.
4) Finally in the last stage grain growth is still predominating. In some systems even exaggerated grain growth can take place forming a very coarse grained structure. This exaggerated grain growth effectively stops the densification as it isolates most of the pores inside the grains.

As diffusion is involved in these sintering processes they are generally thermally activated each with a specific activation energy. As will be demonstrated in the next sections it is possible to determine the activation energies as well as the controlling mechanism for the processes involved in stages 1 and 2 with SID and related techniques.

5.3.2. SINTERING KINETICS

First Stage Kinetics. For each "isothermal step" in the initial sintering stage the densification can be considered to follow the expression [21]:

$$y = \Delta l/l_0 = (K(T)t)^n \tag{15}$$

where

$$K(T) = A\gamma\Omega D/kTr^P \tag{16}$$

In these equations:

l_0	=	initial sample length at the start of sintering
K(T)	=	Arrhenius consatnt
D	=	diffusion coefficient (cm/sec)
r	=	particle radius (cm)
γ	=	surface tension (erg/cm^2)
Ω	=	vacancy volume (cm^3)
A,n,p	=	constants depending on the sintering mechanism
k	=	Boltzmann constant

As the initial sintering stage extends over several steps the time variable must be removed from this equation before it can be used directly. This can be done by differentiation, which gives:

$$y(dot) = (K(T)/N)y^{-(N-1)} \tag{17}$$

where y(dot) is the shrinkage rate and N = 1/n.

This differential equation can also be expressed as:

$$\ln y(dot) = \ln (K(T)/N) - (N-1) \ln y \tag{18}$$

and plotting ln y(dot) versus ln y, both of which can be determined from the experimental data, a straight line will be obtained for each isothermal step. For these steps the value of N and thus n can be determined from the slope and K(T) from the intercept with the ln y(dot) axis.

Taking the logarithm on both sides of the Arrhenius expression:

$$K(T) = A \exp(Q/RT) \tag{19}$$

where A is the pre-exponential factor and E_a the activation energy, gives

$$\ln(K(T)) = \ln A - E_a/RT \tag{20}$$

and both A and E_a can thus be determined from a standard Arrhenius plot by plotting the ln(K(T)) found for the individual isothermal steps versus the 1/T values for these steps.

Finally it will be noted the diffusion coefficient for the slowest diffusing specie, which generally is the cation, also can be determined by this technique once the value of K(T) is known – see Eq. (16).

First Stage Sintering Studies

CeO_2. The dilatometric measurements on CeO_2 compacts were performed in a horizaontal dilatometer (Netzsch Dilatometer type 402E) in an oxygen atmosphere. The sampel holder used was made of Al_2O_3. The cylindrical compacts were prepared by pressing CeO_2 powder (99.9% purity) at 3 t/cm^2. This powder had already been ball-milled for 50 h, which is assumed to produce nearly spherical particles of uniform size and which make the use of Eq.(15) reasonable. The green density of these compacts was 61% TD and the surface area of the milled powder was 4.59 m^2/g.

The shrinkage curves recorded for the isothermal steps during the first stage sintering of the CeO_2 compacts were fitted polynominally and the shrinkage rate (y(dot)) was calculated as a function of the shrinkage from the expression obtained. Plotting the logarithm of these data a set of straight and parallel lines shown in Figure 5-11 according to Eq. (18)

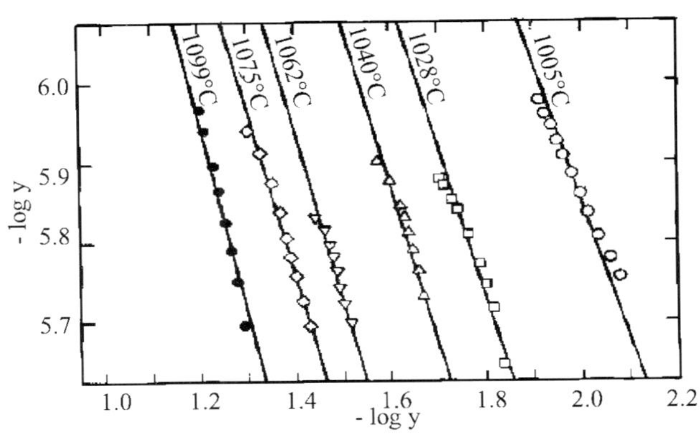

Figure 5-11. Shrinkage rate (y(dot)) versus shrinkage for isothermal steps during sintering of CeO_2.

From the slope of these lines an average value of n = 0.33 was calculated using Eq. (18), which indicates that the shrinkage kinetics for the initial sintering stage of CeO_2 is controlled by "grain boundary diffusion [22,23] A similar conclusion was reached in sintering studies on similar oxides with the fluorite structure [24,26,27].

The ln(K(T)) values could then be determined from the intercepts of the lines as shown above. Plotting these against 1/T gave the straight line shown in Figure 5-12, which in fact confirm that the is thermal steps considered all are governed by the same process.

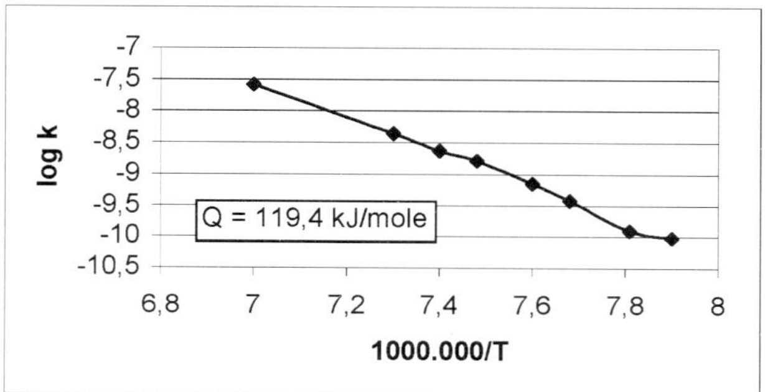

Figure 5-12. Arrhenius constant (K(T)) versus reciprocal temperature (1/T) for sintering of CeO_2 compact.

The activation energy calculated from the slope of this line were E_a = 119.4 kcal/mole (5.2 eV) which correspond well with activation energies reported for cation diffusion for other fluorite oxides [28,29]. At the time of publication of this work activation energies for Ce-diffusion in CeO_2 had not previously been published.

Finally the diffusion coefficients could be determined from the K(T) values obtained in this work using Eq. (16). The values obtained ranged between 8.5×10^{-17} and 3.1×10^{-15} cm^2/sec for the temperature range covered in this measurement (1005–1099°C) which are comparable with data obtained in studies of other fluorite structure oxides [24].

Yttria Partially Stabilized Zirconia. Another important type of engineering ceramics is the so-called transformation toughened zirconias, which are well known for the superior mecnaical properties as well for their excellent electrical properties.

In a recent investigation [30] of the initial sintering stage of these oxides, two types of powders were used: TZ3Y (2.86 mole% of Y_2O_3) and TZ3YA (2.77 mole% Y_2O_3 and 0.1 wt% Al_2O_3) – both powders from Tosoh (Japan).

The logarithmic plot of the shrinkage rate y(dot) versus shrinkage y for TZ3Y is shown in Figure 5-13.

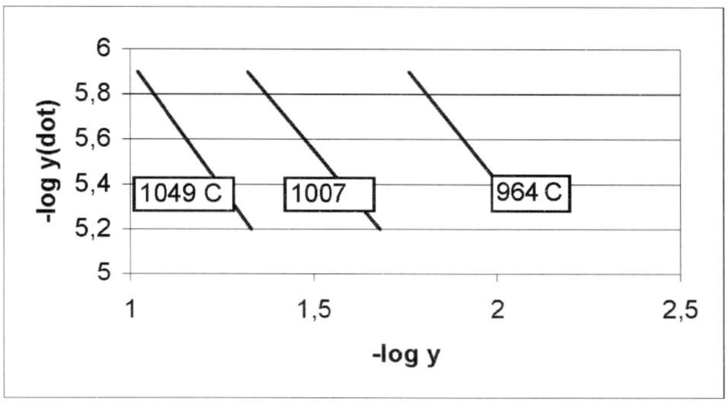

Figure 5-13. Log y(dot) versus log y for TZ3YA compact.

From the slopes the exponent n = 0.31 was calculated, indicating that the grain boundary diffusion is also the controlling mechanism for the first sintering stage for this material. The same value for n was also observed for TZ3YA. For both materials this n-value correspond well with that reported for other compositions of Y-doped zirconia [31].

Figure 5-14 shows the Arrhenius plot of log K(T) vs. 1/T for TZ3Y:

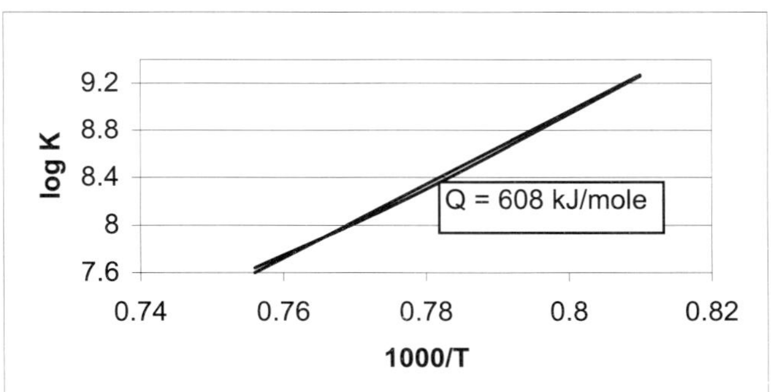

Figure 5-14. Arrhenius plot for TZ3Y.

A nice straight line was obtained from which an activation energy of 608 kJ/mole was obtained. The similar plot for TZ3YA gave a comparable value of 610 kJ/mole. That the activation energies for the two materials are quite similar might be due the similar content of the trivalent dopant in the two powders. Furthermore these values are much higher than for anion diffusion, which suggest that it is the cation diffusion which is the rate controlling also in these oxide systems. Finally these values are also comparable to those deduced from the creep data for grain boundary diffusion in TZ3Y [32], which again support the conclusion of grain boundary diffusion as the rate-controlling mechanism.

UO_2. The SID technique was also used in a study of the kinetics of the first sintering stage of UO_2 compacts as a function of the oxygen pressure [25]. Again the controling mechanism was found to be grain boundary diffusion of uranium ions with an activation energy of 234 kJ/mole in H_2 – 10% CO_2 compared to 390 kJ/mole in pure H_2. These activation energies compares well with litterature data for UO_2 [33].

Intermediate Stage Kinetics

The general equation for the initial sintering stage (Eq. (15)) can also be used to describe the three-dimensional isothermal sintering process by inserting the relative volumetric shrinkage instead of the linear shrinkage [34], which gives:

$$(V_0 - V_t)/(V_t - V_f) = [K(T)\,(t - t_0)]^n \tag{21}$$

where V_0 is the initial volume, V_t the volume at time t and V_f the volume of the fully dense specimen, K(T) the Arrhenius constant and n a constant depending on the mechanism. This equation has been successfully used for describing liquid-phase sintering of alloys [35] and solid state sintering of some metals [36].

From this equation the normalized rate equation has been developed [37]:

$$f(Y) = dY/dt = nK(T)Y(1 - Y)[(1 - Y)/Y]^{1/n} \tag{22}$$

where Y is the fractional densification. Rearrangement and taking the logarithms gives:

$$\ln[f(Y)/Y(1 - Y)] = \ln[nK(T)] + 1/n\,\ln[(1 - Y)/Y)] \tag{23}$$

and plotting $\ln[f(Y)/Y(1 - Y)]$ vs. $\ln[(1-Y)/Y]$ should thus give a straight line from which n can be determined from the slope and K(T) from the intercept.

The activation energy can thus easily be determined by plotting ln K(T) vs. 1/T in a standard Arrhenius plot.

The activation energies can however also be determined directly irrespective of the sintering mechanism by the so-called temperature jump-method. Using SID this can be done using two methods:

1) The sample is quickly heated between the isothermal steps by which it can be assumed that f(Y) is constant before and after the temperature jump, and
2) By changing the rate in a constant shrinkage rate measurement, which produces a change in the temperature level.

In both methods the activation energy can be calculated from the equation [37]:

$$(dY/dt)_1/(dY/dt)_2 = k_1/k_2 = \exp[-E_a/R(1/T_2 - 1/T_1)] \quad (24)$$

where k_1, k_2 are the rates before and after the temperature change, E_a the activation energy and T_1 and T_2 the two temperature levels. The application of this equation to determine activation energies for different types of compacts will be demonstrated in the next section.

Yttria Partially Stabilized Zirconia

TZ3YA

A study of the densification kinetics was performed using TZ3YA (transformation toughened zirconia with 5.14 wt% Y_2O_3, 3.6 wt% acrylic binder and 0.03 wt% Al_2O_3) powder pressed at three different pressures to three different green densities (39.24, 43.81 and 48.67%TD).

Straight lines were obtained for each of the isothermal steps by plotting ln[f(Y)/Y(1 – Y)] vs. ln[(1 – Y)/Y)] as predicted from Eq. (xx) and as shown in Figure 5-15.

The n-values obtained from this plot as well as from the similar plots for the two other samples (TZ3YA-05, TZ3YA-07) were somewhat smaller than the values obtained by using the fundamental equation (Eq. (23)) for the first stage discussed previously. The reason is believed to be due to the different development in the microstructure morphology in these samples, which actually was revealed by a SEM examination. The essentially constant n-values in the low temperature region (900–1050°C) however seems to indicate that the controlling mechanism in this temperature range is neck-growth, whereas the more variable n-values at higher temperatures might indicate a multi-mechanism in the morphology development.

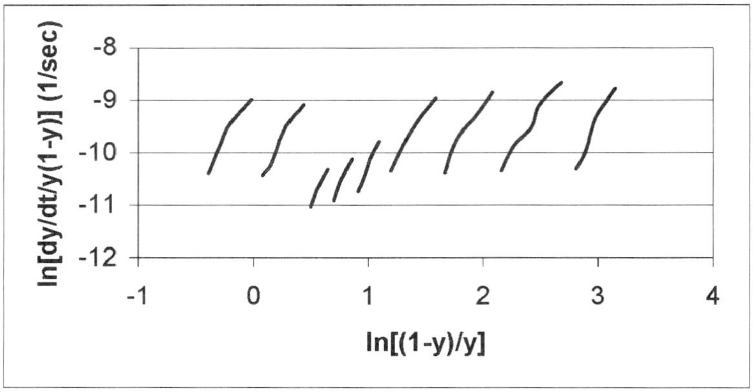

Figure 5-15. Plots for isothermal steps of TZ3YA-03.

This is also demonstrated in the Arrhenius plot (Fig. 5-16) constructed from the lnK(T) values determined from Figure 5-15 and the similar plots for the other samples.

The average activation energies determined for the three samples were:

TZ3YA-03 (39.24%TD) 632.8 kJ/mole
TZ3YA-05 (43.81%TD) 620.4 kJ/mole
TZ3YA-07 (48.67%TD) 592.3 kJ/mole

indicating that increasing green density gives a decrease in the activation energy. These activation energies agree well with the energies determined by Ali et al. [30] (610 kJ/mole) on TZ3YA using the fundamental neck-growth equation. The extensive low temperature densification thus seem to follow this mechanism.

Figure 5-17 shows the activation energies obtained using the temperature jump method assuming that the temperature changes so quickly from one isothermal step to the next that the value of f(Y) is unchanged.

As noted from this plot, the activation energies determined by the jump method are slightly smaller, but still consistent, than the values determined by the more fundamentally correct method.

TZ2.5Y

In another study on the sintering behavior of yttria partially stabilized zirconia, but this time on TZ2.5Y (2.5 mole% of Y_2O_3), the activation energies were determined using the constant rate shrinkage technique. Figure 5-18 shows the shrinkage vs. time and temperature vs. time curves obtained. The figure clearly shows the temperature jumps each time the rate is changed.

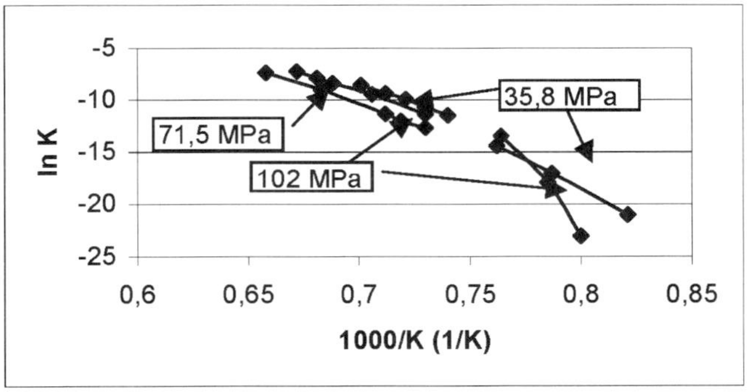

Figure 5-16. Arrhenius plot for specimens TZ3YA-03, TZ3YA-05 and TZ3YA-07 pressed at respectively 35,8, 71,5 and 102 MPa.

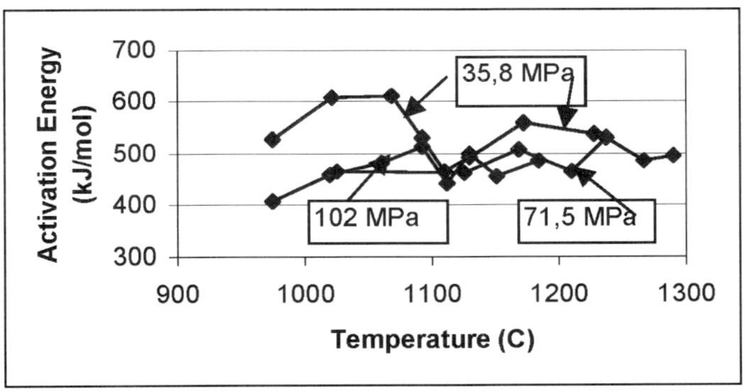

Figure 5-17. Activation energies vs. temperature using the temperature jump method for the three TZ3YA specimens.

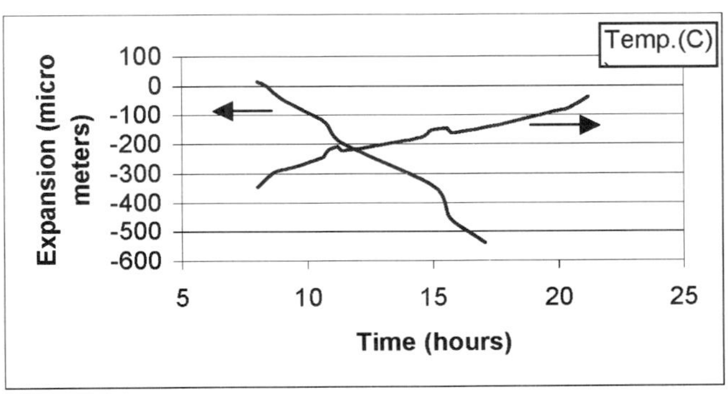

Figure 5-18. Constant shrinkage rate measurement on TZ2.5Y. The change in rate produces a jump in the temperature.

Table 5-3. Activation energies and characteristic data for constant rate shinkage jump measurement on TZ2.5Y

Jump	y(dot)$_i$ (10^{-6})	y(dot)$_{i+1}$(10^{-6})	T_i(°C)	T_{i+1}(°C)	Q (kJ/mol)
1	–4,106	–8,390	1093	1131	300
2	–8,441	–3,914	1143	1127	505
3	–4,199	–8,411	1194	1223	430
4	–8,411	–4,208	1245	1223	605

The characteristic rates and temperature changes as well as the activation energies calculated using Eq. (17) are given Table 5-6.

Compared to literature data, especially the activation energy calculated for the first jump seem to be too small – these vary between 500 and 650 kJ/mole for the same material [38]. The high values observed at higher temperatures probably indicates that grain growth is a contributing factor as proposed by Theunissen [38]. The lower activation energy found for the first jump is however too low compared to the values found for other systems using the SID technique. One reason could be that the power supply to the dilatometer is not high enough to give a sufficienly quick temperature jump.

Al_2O_3

Finally the sintering of alumina was recently studied by SID but using the intermediate stage kinetics. The measurements were performed on compacts prepared by extrusion of an α-Al_2O_3 paste with addition of a kaolin and organic additives as sintering aid [34]. Plotting the values of ln[(dY/dt)/(Y(1 – Y)] versus the values of ln[(1 – Y)/Y] calculated from the experimental data again resulted in a series of straight parallel lines from both the n-values and the ln K(T) values could be obtained as discussed previously.

For the temperature range 1100–1400°C the n-values obtained ranged between 0.213 and 0.251, with the smallest value at the highest temperature (1400°C) and with an average value of 0.232. These values correspond well with those obtained in the similar study on TZ3YA discussed in the previous section, but these n-values were again somewhat smaller than the values obtained by using the fundamental equation for the first stage. The reason in this case is probably the presence of the sintering aid which increases the vacancy flux both during the first nech growth stage and grain boundary diffusion at the intermediate stages. This is also indicated by the somewhat smaller activation energy, 415.5 kJ/mole in the temperature range 1200–1400°C, found in this study.

5.4. SCTA and Material Synthesis (J. Criado)

In the previous chapters it has been described that SCTA methods allow an effective control of both the reaction rate and the partial pressure exerted in the close vicinity of the sample by the gases generated or consumed in the reaction, which permits to minimize the influence of heat and mass transfer phenomena on the forward reaction This ability of SCTA for controlling the surrounding atmosphere and minimizing the thermal gradients across the sample are extending the applications of these techniques for synthesizing materials with controlled texture and structure. The first publications concerning the use of SCTA for the synthesis of inorganic oxides were authored by Rouquerol et al., who used Constant Rate Thermal Analysis (CRTA) (essentially in the form of Constant Rate Evolved Gas Detection (CR-EGD)) to prepare aluminas [40–42] and beryllias [43] with highly controlled pore-size, from the corresponding hydroxide precursors. The application of other forms of CR-EGD to the preparation of aluminas was also examined by Stacey [44,45] and Barnes and Parker [46] as described in Chapter 6 of this book

The Sample Controlled Thermal Analysis was also used for the synthesis of α-Fe_2O_3 (hematite) from the thermal decomposition under vacuum of α-FeO(OH) (goethite) samples with acicular shaped particles [47,48]. An independent control of both a constant reaction rate and a constant pressure of the residual water vapour was allowed by the device used. It was demonstrated that the porosity can be tailored through a proper control of both the reaction rate and the residual pressure of the water vapour generated from the thermal decomposition of the precursor. Slit pore channels oriented along the c-lattice axis (the long axis of the particle) were formed at very low water vapour pressures as shown in Figure 5-19. It was reported that when the water vapour pressure was increased isolated round pores were progressively formed and that the size of these was increasing as a function if the pressure as shown in Figure 5-19. A diminution of the water vapour pressure during the thermal decomposition of goethite strongly promoted the increase of the specific surface of the hematite obtained as the final product [47,48]. A similar behaviour was reported for the textural properties of the γ-Fe_2O_3 (maghemite) obtained by controlling both the rate and the water vapour pressure during the thermal decomposition of the γ-FeOOH (lepidocrocite) precursor by means of the SCRT method [49].

The SCTA method has been also applied to the synthesis of ferroelectric ceramics based on $BaTiO_3$ prepared by thermal decomposition under vacuum of barium tytanil oxalates and citrates [50–53], respectively. The decomposition of the precursors was carried out by controlling the temperature in such a way that both the decomposition rate and the partial pressure were maintained constant at

(a)

(b)

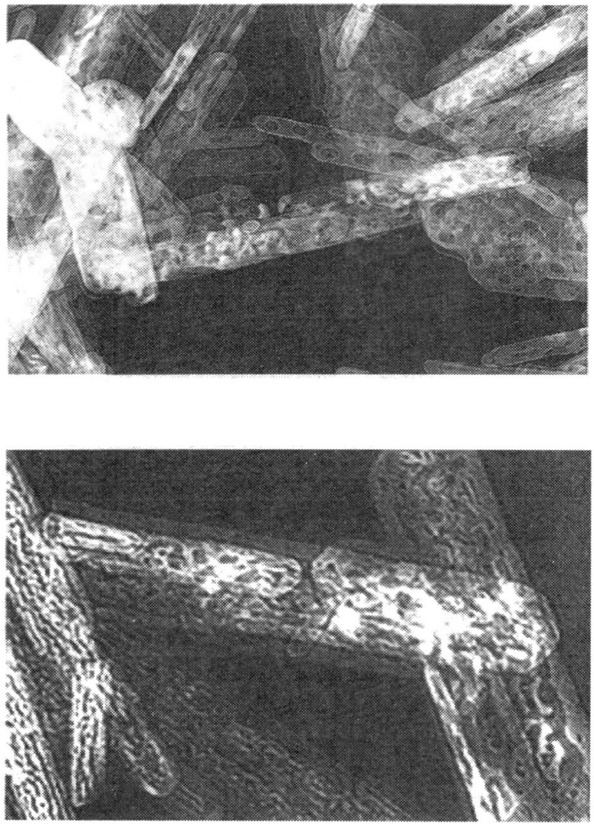

Figure 5-19. TEM micrographs corresponding to hematite sample prepared by thermal decomposition of goethite under CRTA conditions: a) P = 5.5x10^{-5} torr and = 3.3x10-3 min^{-1} b) P = 8.3 torr and C = 7.6x10^{-4} min^{-1}.

the value previously selected by the user. It was shown that the crystal size and the stabilization of the cubic phase with regards to the tetragonal phase of $BaTiO_3$ was strongly depending on the partial pressure of the gases produced in the decomposition of the precursors in the range from 10^{-5} to 10 mbar. The crystal size of barium titanate decreased by decreasing the constant pressure used. The diminution of the particle favoured the stabilization of the cubic phase.

The SCTA method allowed to obtain nanocrystals of $BaTiO_3$ from mixed oxalate and citrate precursors, which was not possible when these precursors were decomposed under conventional isothermal or rising temperature conditions. The proper control of the pressure and the reaction rate by means of the CRTA method during the thermal decomposition of acicular shaped particles of barium titanyl citrate permitted to obtain fibers of $BaTiO_3$ composed by nanocrystals of this compound as shown in Figure 5-20 b and c. The thermal decomposition of the same citrate precursor under conventional isothermal decomposition lead to round powder particles of $BaTiO_3$ and did not permit formation of fibers. CRTA methods have been also reported to improve the densification [55–58] and to suppress exaggerated grain growth during sintering of nanosized barium titanate [59]. The CRTA method has been used as well for controlling the densification and the microstructural development of ceramic oxides [60–70] in order to improve their electrical or optical properties.

SCTA methods have proved to be a powerful tool for optimizing the organic binders removal during the compaction of ceramics without crack formation. Sorensen [71] developed a SIA method for removing the binder in such a way that the crack formation was minimized. The CRTA method has been also successfully used for this purpose [72–76]. Speyer et al. have reported a very nice example of the application of CRTA to the manufacture of multilayers capacitors (MLC) and multilayer actuators (MLA). The ferroelectric ceramics are mainly used for manufacturing MLC or in general MLA, which consists of layers of ceramics separated by electrode metal layers. The manufacturing process implies stacking the different layer by a tape casting technology, generally using an organic binder, followed by a co-firing process. The elimination of the binder during the co-firing treatment is the main problem in the manufacturing of these devices due to internal cracking, resulting from imperfections formed during binder extraction and sintering. Thus, the process of elimination of the binder is a rather cumbersome process that very often takes several weeks [72]. Speyer et al. [76] have shown that this problem can be overcome by controlling the temperature in such a way that the decomposition rate of the binder was maintained constant by using the CRTA method. According to these authors, MLAs can be obtained without imperfections if a low enough decomposition rate is selected. This is illustrated in Figure 5-21 that shows the optical microscope pictures of the polished surface of cofired MLAs obtained under different constant rate binder loss percentages. It is clearly shown that the delaminating damage is dramatically reduced by decreasing the rate of elimination of the binder.

The use of SCTA methods to control both the reaction rate and the partial pressure of the CO generated in the carbothermal reduction of silica allowed a successful control of the phase composition and the crystal and particle size of Si_3N_4 [77–79] which would not have been possible using conventional heating schedules. This is illustrated [80] in Figure 5-22 which shows the SEM micrographs of two samples of Si_3N_4 annealed at 1450°C under a flow of 95% nitrogen + 5% hydrogen for 5 hours, but after being synthesized from carbothermal nitridation of silica at a constant rate $d\alpha/dt = 1.1.10^{-3}$ min^{-1} and two different residual pressures of the CO generated in the reaction. These SEM pictures point out that the sample a) obtained at the lowest constant residual pressure of CO consists of a mixture of β-Si_3N_4 ribbons and small hexagonal crystallites of α-Si_3N_4 while the sample b) obtained at the higher residual CO pressure consists of hexagonal crystallites of pure α-Si_3N_4 of homogeneous size. The synthesis of SiC wiskers by SCTA has been also described in [80].

Figure 5-20. (a) Scanning Electron Micrograph (SEM) of the original barium titanium citrate (BTC). (b) SEM of the $BaTiO_3$ sample obtained by annealing at 700°C for 5 hours the BTC. (c) SEM of the $BaTiO_3$ obtained under CRTA conditions ($P = 10^{-5}$ torr). The sample was allowed to reach 700°C and was hold at this temperature for 5 hours (d) Transmission Electron Micrograph (TEM) of the same sample as in c.

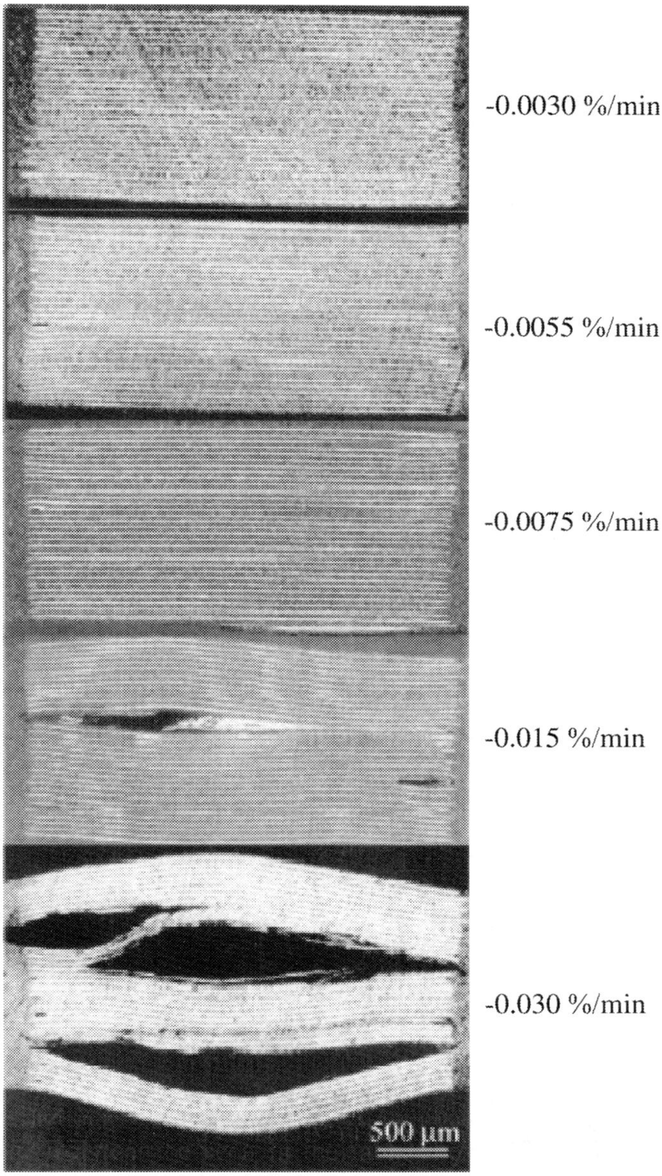

Figure 5-21. Appearance of coarse grained MLAs after varying SCRT treatment, after mounting on epoxy and grinding from the surface, using an optical microscope.

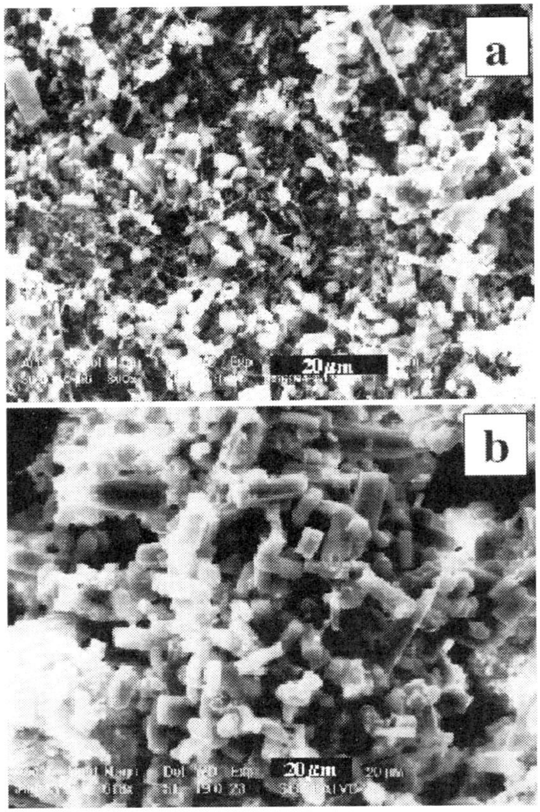

Figure 5-22. SEM micrographies of silicon nitride obtained from carbothermal reduction of silica: a) CO = 0.0002 atm and C = $1.1.10^{-3}$ min^{-1} and b) CO = 0.01 atm and C = $1.1.10^{-3} min^{-1}$.

The SCTA method was recently applied to the thermal curing of concretes [81]. It was shown that the control of the dehydration rate in such a way as to maintain very reduced water vapour partial pressure in the close vicinity of the samples hinder the crystallization of xonotlite and in general de crystallization of hydrates which promotes the strengthening of concretes. Application of SCTA, in the form of CR-EGD, was shown to allow the control of the interfiber porosity and to eliminate the structural microporosity of sepiolites [82]. It proved to be a powerful method to synthesize oxides with tailored specific surface areas [83–85] and also to totally separate the dehydration and denitration steps in the thermolysis of hexahydrated uranyle nitrate: these were shown to be essential conditions to obtain highly divided and reactive UO2 [86–88].

References

1. F. Chen, O.T. Sørensen, G. Meng and D. Peng, J. Thermal Analysis 53 (1998) 397–410.
2. M. Stockenhuber, H. Mayer and J.A. Lercher. J. Am. Ceram. Soc., 76 (1993) 1185.
3. F.L. Chen, P. Wang, O.T. Sørensen, G.Y. Meng and D.K. Peng, J. Mater. Chem., 7 (1997) 1533.
4. I. Aboltina, R. Ramata and I. Brante, Ferroelectrics, 141 (1993) 277.
5. V.B. Reddy and P.N. Kaushik, Thermochim Acta, 83 (19985) 347.
6. K.K. Kelly, Bur. Mines. Bull, No. 371 (1934).
7. K.K. Kelly and C.T. Anderson, Bur. Mines Bull No 384 (1935).
8. A. Finkelstein, Ber., 39(1906) 1585.
9. L. Hackspill and G. Wolf, Compt. Rendu, 204 (1937) 1820.
10. E.H. Baker, J. Chem. Soc., (1964) 699.
11. C.H. Bamford, C.F.H. Tipper: Reactions in the solid state. Comprehenseive Chemical Kinetics., vol. 22, Elsevier 1980.
12. T.P. Bagchi and P.K. Sen, Thermochim Acta., 51 (1981) 175.
13. A.W. Coats and J.P. Redfern, Nature, 201 (1964) 68.
14. E.S. Freeman and B. Carroll, J. Phys. Chem., 62 (1958)394.
15. U. Patnik and J. Muralidhar, Thermochim. Acta, 274 (1972) 31.
16. M.D. Judd and M.I. Pope, J. Thermal. Anal., 4(1972) 31.
17. T. Kudo and H. Obayashi, J. electrochem. Soc., 122(1975)142–147.
18. J.J. Bentzen, P.L. Husum and O. Toft Sørensen, in High Tech ceramics (Ed. P. Vincenzini) Elsevier Science Publishers, 1987, 385–398.
19. P.L. Husum and O. Toft Sørensen, Thermochim Acta, 114 (1987) 131–138.
20. O. Toft Sørensen, J. Thermal Anal., 38 (1992) 228.
21. M. El- Sayed and O. Toft Sørensen. Initial Sintering Stage Kinetics of CeO_2 studied by Stepwise Isothermal Dilatometry. Risø-R-518. February 1985. Risø National Laboratory, Denmark.
22. R.L. Coble,. J. Am. Ceram. Soc., 41, 1958, 55–62.
23. W.S. Coblenze et al., In "Sintering Processes, Materials Science and Research" (Ed. G.C. Kuczynski) 13, 1980.
24. M. El-Sayed Ali, O. Toft Sørensen and L. Hälldahl, In "Thermal Analysis, proceedings of the seventh International Conference on Thermal Analysis". (Ed. Ed. B. Miller), Wiley, NY 1982, 344–349.
25. M. El-Sayed Ali, O. Toft Sørensen and L. Hälldahl, J. Therm. Anal., 25, 1982, 175–180.
26. J.J. Bacman and C. Cizeron, J. Nucl. Mater., 33, 1969, 271–285.
27. M. El-Sayed Ali and R. Lorenzelli, J. Nucl. Mater., 87, 1979, 90–96.
28. H.J. Matzke, In "Nonstoichiometric Oxides" (Ed. O. Toft Sørensen), Academic Press, NY, 1980, 155–232.
29. P. Kofstad, In "Non-stoichiometry, Diffusion and Electrical Conductivity in Binay metal Oxides", Wiley, NY 1973, 276–283.

30. M- El-Sayed Ali, S. El-Houte and O. Toft Sørensen, Interceram., 40(4), 1994, 248–250.
31. W.H. Rhodes, J. Am. Ceram. Soc., 64(1), 1981, 19–22.
32. Y. Okamot, J. Leujt, Y. Yamada, K. Hayashi and T. Nishikama, In "The 3rd International Conference on the Science and Technology of Zirconia", Extended Abstracts, Zirconia, Tokyo, 1986.
33. Hj. Matzke, in "Nonstoichiometric Oxides" (Ed. O. Toft Sørensen) Academic Press 1981, 155–232
34. Huan-ting Wang, Xing-qin Liu, Fang-lim Chen, Guang-yao Meng and O. Toft Sørensen, J. Am. Ceram Soc. 81(3) (1998) 81–84.
35. S. Mkipirti, in Powder metallurgy (Ed. W Leszynski) Interscience New York 1996, 97.
36. H.E. Exner and G. Petzov, in Sintering Processes (Ed. G.C. kuczynski) Plenum Press, NY 1980.
37. G.Y. Meng and O. Toft Sørensen, "Kinetic Analysis on Low Temperature Sintering Proicess for Y-TZP Ceramics" in Advanced Structural Materials (Ed. Y Han) Elsevier Science Publishers, Amsterdam, vol 2, 1991, 369–374.
38. G. Theunissen, Microstructure, Fracture Toughness and Strength of (ultra) Fine Grained Tetragonal Zirconia Ceramics, Thesis, 1991.
39. T. Arii, K. Terayama and N. Fujii, J. Therm. Anal., 47(1996) 1649–1661.
40. J. Rouquerol and M. Ganteaume; J. Thermal Anal., 11 (1977) 201.
41. J. Rouquerol, F. Rouquerol and M. Ganteaume; J. Catal. 36 (1975) 99
42. J. Rouquerol, F. Rouquerol and M. Ganteaume, J. Catal., 57 (1979) 222.
43. F. Rouquerol, J. Rouquerol and B. Imelik, In Principles and applications of pore structural characterization, ed. by J.M. Haynes and P. Rossi-Doria, Bristol, Arrowsmith (1985) 213.
44. M.H. Stacey, Anal. Proc., 22 (1985) 242.
45. M.H. Stacey, Langmuir, 3 (1987) 681.
46. P.A. Barnes and G.M.B. Parkes, Preparation of Catalysts VI (Scientific Bases for the Preparation of Catalysts); (G. Pocelet et al.Eds.) Elsevier, Amsterdam (Holland) 1995.
47. L.A. Pérez-Maqueda, J.M. Criado, C. Real, J. Subrt and J. Bohácek, J. Mater. Chem. 9 (1999) 1839.
48. L.A. Pérez-Maqueda, J.M. Criado, J. Subrt and C. Real, Catal. Letters, 60 (1999) 151.
49. G.S. Chopra, C. Real, M.D. Alcalá, L.A. Pérez-Maqueda, J. Subrt and J.M. Criado, Chem. Mat., 11 (1999) 1128.
50. J.M. Criado, F.J. Gotor, C. Real, F. Jiménez, S. Ramos and J. Del Cerro; Ferroelectrics, 115 (1991) 43.
51. J.M. Criado, M.J. Diánez, F. Gotor, C. Real, M. Mundi, S. Ramos and J. Del Cerro, Ferroelectric Letters, 14 (1992) 79.
52. F.J. Gotor, C. Real, M.J. Diánez and J.M. Criado, J. Solid State Chem., 123 (1996) 301

53. F.J. Gotor, L.A. Pérez-Maqueda and J. M. Criado, J. Eur. Ceram. Soc., 22 (2002) 2227.
54. L.A. Pérez-Maqueda, F. Gotor, M.J. Diánez, C. Real and J.M. Criado, In press.
55. K.S. Meyers, M. Seivastava and R.F. Speyer, Proc. SPIE-Int. Soc. Opt. Eng. 3330 (1998).
56. A.V. Ragulya, Nanostruct. Mater., 10 (1998) 349.
57. A.V. Ragulya and A.V. Polotay, Ferroelectrics, 254, (2001) 41.
58. A.I. Bykov, A.V. Polotay, A.V. Ragulya and V.V. Skoeokhod, Powder Metall. Met. Ceram., 39 (2001) 395.
59. K.S. Meyers and R.F. Speyer, Mater. Res. Soc. Symp. Proc., 547 (1999) 115
60. G. Agarwall, R.F. Speyer and W.S. Hakenberger, J. Mater. Res., 11 (1996) 671.
61. G. Agarwall and R.F. Speyer, Mater. Res. Soc. Symp. Proc., 431 (1996) 427.
62. A.V. Ragulya, V.V. Skorokhod and M.G. Burenkov, Key Eng. Mater., 132–136, (1997) 674.
63. G. Agarwall and R.F. Speyer, J. Mater. Res., 12 (1997) 2447.
64. D. Hudda, M.A. El Baradie, M.S.J. Hashmi and R. Puyane, J. Mater. Sci., 33 (1998) 271.
65. J. Zimmer, F. Roether, K. Jaenicke-Rossler and G. Leither, Adv. Sci. Tecnol., 14 (1999) 693.
66. G. Gilde, P.A. Patel and M. Patterson, Proc. SPIE-Int. Soc. Opt. Eng., 3705 (1999) 94.
67. K. Maca, H. Hadraba and J. Cihlar, EUROMAT 99, Biannu. Meet. Eur. Mater. Soc. (FEMS) 12 (2000) 161.
68. O.B. Zgalat-Lozynskyy, A.V. Ragulya and M. Herrmann, NATO Science Series, II: Mathematics, Physics and Chemistry, 16 (2001) 161.
69. Y. Masuda and H. Satoh, Netsu Sukutey, 28 (2001) 193.
70. T.R.G. Kutty, K.B. Khan, P.V. Hedge, A.K. Sengupta, S. Majumdar and D.S.C. Purushotham, J. Nucl. Mater., 297 (2001) 120.
71. O.T. Sorensen, J. Thermal Anal., 38 (1992) 213.
72. A. Dwivedi and R.F. Speyer, Thermochim. Acta, 247 (1994) 431.
73. W.S. Hackenberger, T.R. Shrout and R.F. Speyer, Sintering Technol. [Conf.] 505 (1996).
74. T. Arii, K. Terayama and N. Fujii, J. Thermal Anal., 47 (1996) 1649.
75. J. Witt, R.F. Speyer and L. Murali, Rev. Sci. Instrum., 68 (1997) 2546.
76. M.Y. Nishimoto, R.F. Speyer and W.S. Hackenberger, J. Mat. Sci., 36 (2001) 2271.
77. M.D. Alcalá, C. Real and J.M. Criado, J. Thermal Anal., 38 (1992) 313.
78. M.D. Alcalá, J.M. Criado and C. Real, Mat. Sci. Forum,383 (2002) 25.
79. M.D. Alcalá, J.M. Criado and C. Real, Adv. Eng. Mater., 4 (2002) 478.
80. C. Real, M.D. Alcalá and J.M. Criado, Solid State Ionics, 95 (1997) 29.
81. A. Feylessoufi, M. Crespin, P. Dion, F. Bergaya, H. Van Damme and P. Richard, Advanced Cement Bases Materials, 6 (1997) 21.

82. Y. Grillet, J.M. Cases, M. Francois, J. Rouquerol and J.E. Piorier, Clays and Clay Min., 36 (1988) 233.
83. P.L. Llewellin, V. Chevrot, J. Regai, O. Cerclier, J. Estienne and F. Rouquerol, Solid State Ionics, 101–103 (1997) 1293.
84. E. Diez, O. Monnereau, L. Tortet, G. Vacquier, P. L. Llewellin and F. Rouquerol, J. Optoelectr. Advan. Mater., 2 (2000) 552.
85. T. Arii, T. Taguchi, A. Kishi, M. Ogawad and Y. Sawada, J. Eur. Ceram. Soc. 22 (2002) 2283.
86. S. Bordère, A. Floreancing, F. Rouquerol and J. Rouquerol, Solid State Ionics 63–65 (1993) 229.
87. S. Bordère, F. Rouquérol, J. Rouquérol, J. Estienne and A. Floreancig; J. Therm. Anal., 36 (1990) 1651–1668.
88. S. Bordère, F. Rouquérol, P.L. Llewellyn and J. Rouquérol, Thermochim. Acta, 282–283 (1996) 1–11.

Chapter 6

SCTA AND ADSORBENTS

P. LLEWELLYN, F. ROUQUEROL and J. ROUQUEROL

MADIREL Laboratory, CNRS-Université de Provence, Marseille, France

6.1. Introduction

The preparation, characterisation and pre-treatment (or "outgassing") of *technological adsorbents* (for gas adsorption, liquid adsorption or heterogeneous catalysis) was among the two first applications of Sample Controlled Thermal Analysis (SCTA) (the other being the study of phase changes in metallic alloys [1]) and it was the topic of most papers published about SCTA in the sixties (with the exception of one paper on sintering [2]). This interest is still as vivid as ever and can be explained by the fragility of the surface state and porous texture of a number of these materials and by the difficulty to obtain highly homogeneous and reproducible adsorbents at the end of a thermal treatment. We shall see, indeed, that SCTA, mainly in the form of Controlled Rate Thermal Analysis (CRTA) can provide a *reproducible* thermal pathway to a given adsorbent state, may this be during *adsorbent preparation* where it is possible to isolate interesting intermediate states, or whilst *outgassing* prior to the adsorption application in question. Such reproducibility is obtained due to the ability to control and thus minimise (if the rate of transformation or outgassing is fixed at a low value) both temperature and pressure gradients within each individual grain as well as in the sample bed. A control of the temperature gradient allows the minimisation of hotspots in the sample bed in case the sample heating auto-accelerates. Let us also notice that the control of the pressure gradients within a sample is interesting from an engineering point of view to avoid "sputtering of the sample" out of the container, especially in the scope of vacuum heat treatments.

6.2. SCTA and Adsorbents Preparation

A thermal step is used in the preparation of most adsorbent powders, especially those which are prepared by the thermal treatment of a precursor (hydroxide, carbonate, oxalate, but also wood or coal ...). This thermal step is critical with respect to basic features of the adsorbent, like its surface chemistry, surface texture (grain and pore size and shape, surface area), surface crystalline structure and, above all, homogeneity. Nevertheless, over years, physical chemists tended to underestimate the importance of this step, which was therefore treated relatively simply, with standard conditions like a temperature ramp followed by a plateau, over a pre-selected time. Such a treatment can provide repeatable conditions provided all details (amount of precursor, air or carrier gas flow, shape of crucibles) are kept unchanged.

A few weaknesses of this approach are that (i) it usually provides heterogeneous products (due to the absence of control of the gradients in the sample during its transformation) so that only part of the adsorbent has the expected properties (ii) it may give rise to complex transformations due to uncontrolled overlap of successive steps (a consequence of heterogeneous heating) and (iii) it is very difficult to model at another scale, like for instance that of a laboratory experiment, so that the understanding of what happens in the heat treatment furnace is unsafe and uneasy.

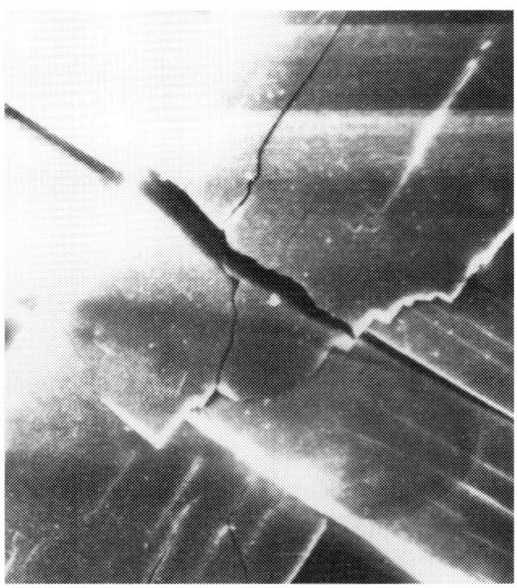

Figure 6-1. Cracking in zeolite crystals: a demonstration of the degradation caused by the lack of control during thermal treatment.

Just as a practical example, Figure 6-1 shows the grain cracking resulting from the non-controlled heat treatment of a zeolite. The interest of SCTA, specially in the form of Controlled Rate Thermal Analysis (CRTA), for the preparation of adsorbents is illustrated hereafter with the case of activated alumina, zeolites and MCM-41 as well as for carbon activation.

6.2.1. ACTIVATED ALUMINA

As we shall see, the preparation of activated alumina by CRTA is a real textbook case. There are several reasons for this: (i) because prior to the use of CRTA, the steps and mechanisms involved were highly controversial, (ii) because this process proved to be extremely sensitive to experimental conditions, so that the close control provided by CRTA was extremely worthwhile, (iii) because of its industrial importance and (iv) because the preparation of this material was certainly the most studied by CRTA, at least in four different laboratories, over 35 years!

Most of the alumina produced in the world is obtained by thermal decomposition of a precursor which is usually gibbsite $Al(OH)_3$, itself prepared industrially, from bauxite ore, by the Bayer process. This thermal decomposition, which is essentially a dehydroxylation accompanied by a number of structural changes, is not a problem when the aim is to prepare the low surface-area α-alumina used to feed the electrolytic cells to produce pure aluminium. Now, when the goal is to produce an activated alumina, to be used as an adsorbent or as a catalyst support, the situation is much more delicate and, for instance, a final product with narrow, pre-determined, pore-size distribution is rarely obtained by standard heating techniques. Also, the heterogeneous conditions resulting from a conventional heating make it very difficult and questionable to find a mechanism for the thermal transformation and for the formation of the porous structure. The many studies carried out on this transformation prior to the sixties show the complexity of the phenomenon, involving a large number of "transition aluminas" [3], with a complex porous structure [4,5] often coexisting but able to follow different paths, depending on the experimental conditions [6]. Hence the idea of studying this transformation by CRTA, as was extensively done, by Rouquerol et al., in the sixties and seventies [7–18]. CRTA was the central technique used in this work and the samples it allowed to prepare in carefully controlled conditions were also studied by IR spectroscopy [7], NMR [9], nitrogen adsorption [8,10,12,13], DSC (either alone or with simultaneous CR-EGD) [6], TG (also, either alone or with simultaneous CR-EGD) [17], transmission or scanning electron microscopy [15] and XRD [15,18]. Part of the interest of this study also lies in the use of two types of gibbsite samples, namely the common, industrial one (whose grains, in the 10–150 μm range, are always poly-crystalline) and also a specially prepared gibbsite made of independent mono-crystalline platelets, 1 μm thick, which will be called hereafter 'fine gibbsite'; moreover, a special batch of even finer mono-crystalline platelets (0.2 μm thick) was used for a number of special experiments.

A first striking result is shown in Figure 6-2, where two thermal analysis curves of identical 1g industrial gibbsite samples are reported [11]. Both were obtained under vacuum (whose value close to the sample was recorded and is reported under the corresponding thermal analysis curve). The bottom curve was obtained by standard TG, with a slow heating rate of 0.5 $Kmin^{-1}$, so that the whole experiment, from 20 to 1000°C, lasted *ca* 33 hours. The main mass loss, which is due to the thermolysis of gibbsite proper, took place between 150 and 300°C. Now, the upper curve was obtained by Controlled Rate Evolved Gas Detection (CR-EGD), with a constant rate of mass loss of 13.5 mgh^{-1}, which resulted in an overall duration of 27 hours. As can be seen, the main mass loss took place here between 170 and 200°C, i.e. with *a temperature span 5 times smaller* than with the conventional TG experiment. The explanation is simply that the control of the rate of transformation is such, in the CR-EGD experiment, that the self-cooling of the sample and the local overpressure of self-generated gas (here, water vapour) are drastically limited as compared with the conventional experiment.

In other words, the 150°C temperature span observed in the latter experiment for the main dehydroxylation step has more to do with the broadly different experimental conditions imposed to the different grains of the sample than with the nature and structure of gibbsite. This explains that it is hopeless to use this type of TG curve to derive any kinetic parameter and any mechanism for this thermolysis.

A major interest of CR-EGD is not only of ensuring a controlled dehydroxylation rate (here, it is chosen constant), but is also of allowing to keep at a pre-determined value the residual water vapour pressure in the vicinity of the sample. This feature was taken advantage of in order to see and understand the influence of vapour pressure on the development of the porous structure of alumina during the CRTA treatment of gibbsite. The results were especially meaningful, as illustrated in Figure 6-3 where we report, the nitrogen BET specific surface area *vs.* temperature at which the CRTA treatment (carried out under a constant residual vapour pressure, indicated in mbar on the corresponding curve) is provisionally stopped. Here, the starting material was fine gibbsite.

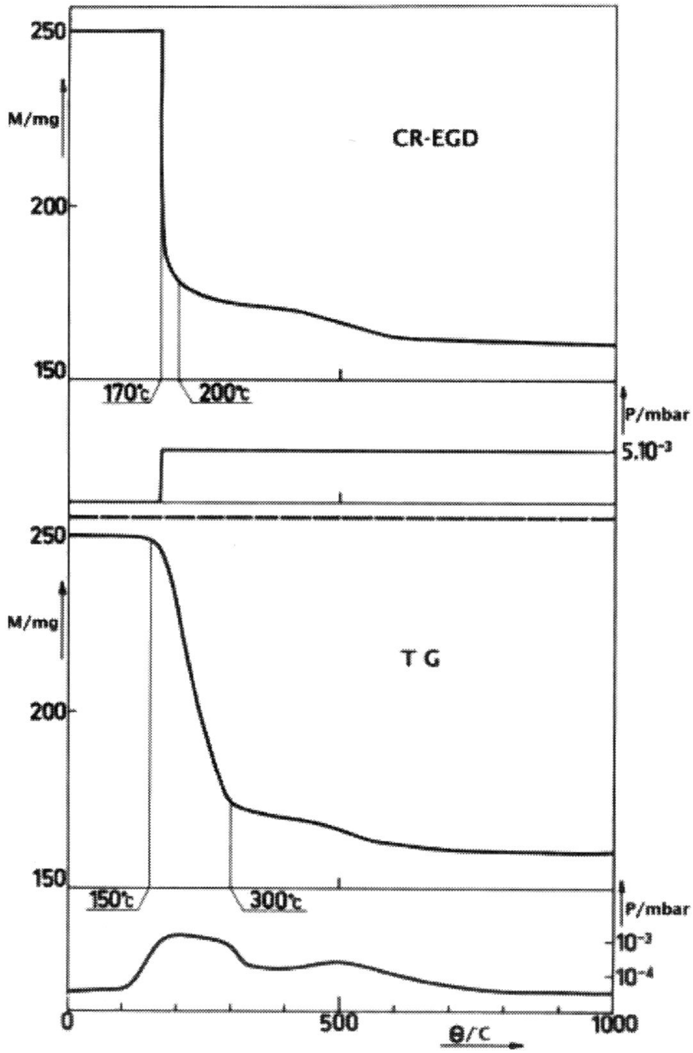

Figure 6-2. CR-EGD curve (top) and TG curve (bottom) for the thermolysis of the same type and mass of gibbsite sample (after [11]).

As can be seen, there is a drastic influence of the residual vapour pressure in the range under 5 mbar. The maximum nitrogen BET surface area reached by the sample is *only 40* m^2g^{-1} when the CRTA experiment takes place under 0.04 mbar vapour pressure, though it *exceeds 400* m^2g^{-1} under 1 or 5 mbar. The general mechanism proposed to explain this observation, together with the

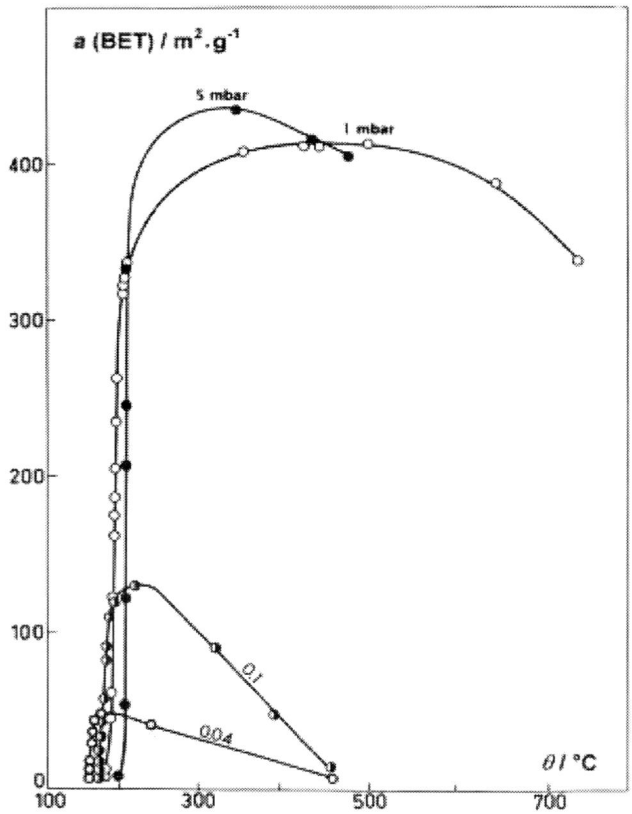

Figure 6-3. Influence of the pressure of self-generated H_2O on the N_2 – BET surface area of an alumina obtained by CRTA treatment of gibbsite $Al(OH)_3$ (after [17]).

many others gathered with help of the complementary techniques, has its starting point in the structure of the gibbsite crystal, which is represented in Figure 6-4. As can be seen, it contains structural channels which are inaccessible to adsorption simply because they are completely lined with hydroxyls (open circles). These channels are all parallel to each other and perpendicular to the basal face (001) of the gibbsite crystal. The "story" of gibbsite dehydroxylation can then be written as follows:

1) From room temperature up to approximately 170°C, a very small mass loss is observed and is only due to the departure of water physisorbed on the outer surface of the gibbsite grains (0.14 m^2g^{-1} for the industrial gibbsite, 15 m^2g^{-1} for the fine one).

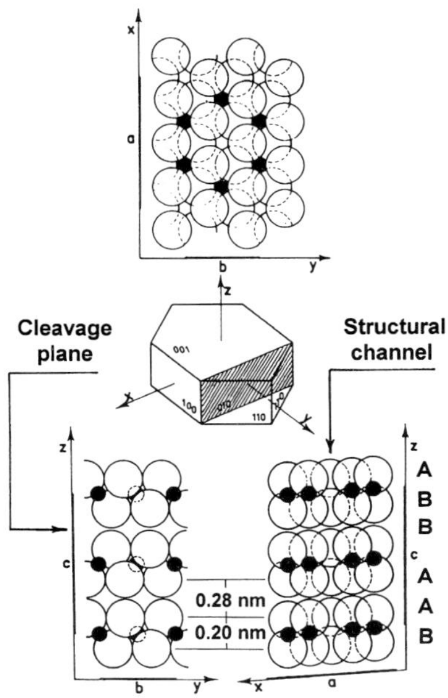

Figure 6-4. The structure of gibbsite, showing the structural channels completely lined with hydroxyls (the open circles) and parallel to the z-axis in the crystal. The black dots represent the location of the aluminum cations (after [16]).

Around 170°C the thermal decomposition proper starts. Now, in the case of industrial gibbsite, no increase in the surface area (i.e. to the surface accessible to the nitrogen molecule) is detected up to *ca* 5% of the full dehydration. Simultaneously, IR spectroscopy shows the appearance and growth of two absorption peaks (3080 and 3260 cm^{-1}) characteristic of a boehmitic phase AlO(OH), whereas NMR shows the appearance and growth of a sharp peak corresponding to mobile molecular water [9]. Microcalorimetry indicates that the differential enthalpy of dehydroxylation is constant, around 50 $kJmol^{-1}$, during the loss of the first 20% of the hydroxyl contents [13]. It is inferred that, in the centre of each polycrystalline crystal, the first reaction to take place is the partial dehydroxylation into boehmite, which is known to happen under hydrothermal conditions. Here, each grain acts like a small autoclave, allowing the build up of the pressure necessary for the transformation, whereas the water, which cannot escape, remains in the state of lacunar free molecular water within the boehmite formed. Such *lacunae* or gaps quite simply result from the density difference between gibbsite (2.42) and boehmite (3.02). This first step of gibbsite dehydroxylation is completely skipped in the case of fine gibbsite dehydrated under a vacuum of 0.02 mbar or lower, but involves *ca* 20% of the starting material in the case of industrial gibbsite dehydrated under 1 mbar. The fact that grain size increases the proportion of boehmite is easily understood by the greater difficulty of the water to escape from the middle of the grain and therefore better hydrothermal conditions. Now, the dependence of this proportion of boehmite on the surrounding vapour pressure even when it is smaller than 5 mbar is, at first sight, more tricky. The explanation proposed lies on the well-known phenomenon of strong adsorption of water in, or at the opening, of ultra-micropores, specially in the case of oxides, and on the possible migration of hydroxyls along the structural channels without any need of emptying them. Under these conditions, adsorption of water at the opening of the channels, which is highly pressure-dependent, partly blocks the evolution of water and imposes a much higher pressure in the centre of the grain.

2) Up to 200°C, the direct and more complete dehydroxylation of the remaining gibbsite (therefore amounting to 70% to 100% of the starting material) into microporous alumina takes place (the term "microporous" is used here with its conventional IUPAC meaning, i.e. to tell that the pores are less than 2 nm wide). This porous alumina is either totally amorphous (if obtained from fine gibbsite under 0.04 mbar or less), or slightly crystallized in the "ρ" form (under 1mbar) or more crystallized, in the "χ" form (under 1 bar). The corresponding differential enthalpy of dehydration is higher than before and, here again, it is constant, at around 77 $kJmol^{-1}$ [13].

The surface area accessible to nitrogen now increases linearly with dehydroxylation. The dehydroxylation temperature, under CRTA conditions, is fairly constant. The higher the residual pressure, the higher the temperature needed to ensure the constant dehydroxylation rate required and also the higher the slope of the surface area curve, as shown in Figure 6-5 (a).

This is quite simply explained by the pre-existing structural channels. At this stage of the dehydroxylation, they start to be emptied from their hydroxyls which are close enough to easily condense and produce water. The emptying starts from the mouth of the channels, i.e. from the basal planes of the crystals. All channels are emptied progressively and simultaneously, so that the micropores now available to nitrogen can be represented as being "drilled" at the same speed, parallel to each other. In case the rate selected for the CRTA experiment is slow enough, the dehydroxylation is not diffusion controlled, so that the rate of the reaction is independent from the depth of the pores: this is why, when carried out by CRTA, this main part of the dehydroxylation is practically isothermal. The pressure dependence of the CRTA curves and of the resulting surface areas tends to show that the dehydroxylation is controlled by a desorption step taking place on the "mouth" of the pores. It is known indeed that water physisorbs quite strongly in "ultramicropores" of molecular size, specially in oxides. It therefore makes sense to consider that a higher pressure makes the desorption more difficult and requests a higher temperature to achieve the desired rate of dehydroxylation. It follows that the aluminium cations gain some mobility which allows an enlargement of the pore size, which is then more accessible to the nitrogen molecule. In other words, a higher water vapour pressure does not change the number of pores, but simply their width. The nitrogen molecule then happens to have the right size to magnify the phenomenon and to provide the dramatic results reported in Figure 6-3. If the above picture is correct, it could be possible, in principle, during a single experiment, to produce pores of different sizes by simply changing the residual pressure; for instance, one could think of first drilling "broad" micropores, fully accessible to nitrogen, by using a vapour pressure of 1 mbar, and then of drilling narrow micropores, in great part inaccessible to nitrogen, by using a vapour pressure of 0.04 mbar. Several experiments of this type were carried out and are presented in Figure 6-5 (b): as expected, the lower slope of the surface area curve in the second part of each experiment (i.e. under 0.04 mbar) indicates that this procedure (which absolutely requires a CRTA control) allows to produce "funnel-shaped" micropores.

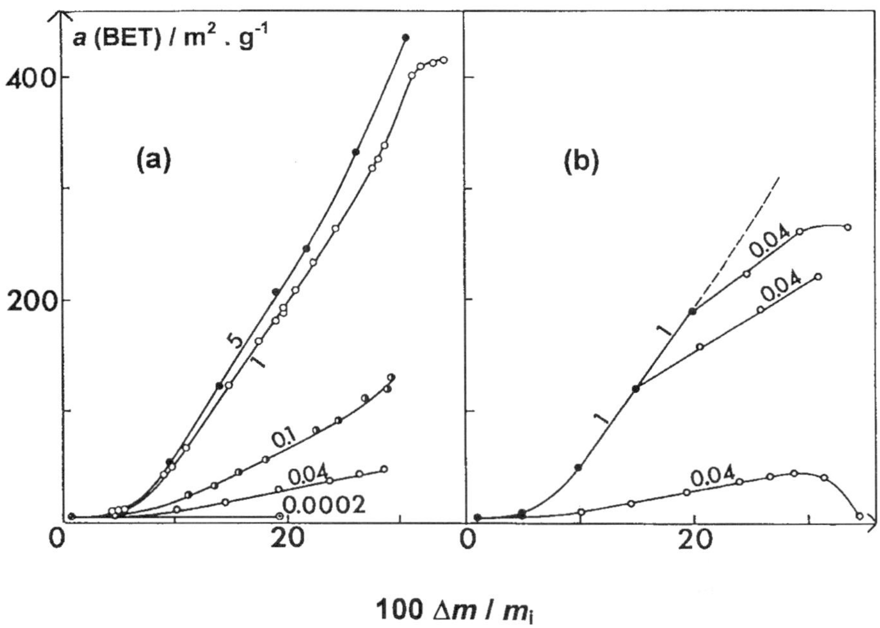

Figure 6-5. (a) same results as in Figure 3, but with BET surface area plotted vs. % of mass lost (b) same type of plot as in (a), but with two successive values of residual pressure (expressed in mbar) during the dehydroxylation (after [16]).

3) Between 200 and 400°C, one observes the progressive dehydroxylation of the porous alumina just formed, together with some shrinkage of the porous structure: the shrinkage is strong for the ultra-microporous aluminas prepared under 0.1 mbar or less (cf Figure 6-3) and which are likely to store an appreciable surface energy in the form of surface defects and curvature; the shrinkage does not exist until 500°C for the sample prepared under 1 mbar (stable enough and slightly crystallized) but it can be observed, again, though moderately, for the alumina prepared under 5 mbar, which is explained by the well-known catalytic role of water towards the shrinkage of oxides.

4) Between 400 and 500°C, one observes the dehydroxylation of the boehmite phase formed in step 1): the IR absorption peaks at 3080 and 3260 cm^{-1} are seen to vanish, together with the NMR narrow peak of molecular, free water, as shown in Figure 6-6. It is worth noting that these NMR spectra were recorded on samples directly prepared under CRTA conditions; they were transferred under vacuum into the NMR test tubes which were eventually sealed, so that there was no risk of rehydration at any stage of the handling.

The broad peak is due to the protons of hydroxyls and also, up to 400°C, of strongly bonded water molecules, whereas the central, narrow peak is ascribed, from 180°C onwards, to the free water molecules enclosed in the boehmite phase. On the starting sample (25°C) it is due to the small amount of water physisorbed on a surface of 0.14 m^2g^{-1}.

This water (and its corresponding narrow NMR peak) disappear on outgassing at 150°C. This thermolysis of the boehmite phase produces a second type of activated alumina but does not give rise to any increase of the total nitrogen BET surface area, probably because it is counterbalanced by some sintering. The amount of boehmite decomposed (and previously formed during step 1) follows directly from the shape of the CR-EGD curve, as shown in Figure 6-7a (influence of the grain size) and 6-7b (influence of the water vapour pressure). It is directly proportional to the length of step d-e (non-existing on curve I, for the smallest grain-size or the lowest pressure, where no boehmite is formed) and corresponds to a yield of 30% and 13% on curves III of Figures 6-7a and 6-7b, respectively [7,14].

5) Above 500°C, as the dehydroxylation proceeds, the surface area, when still important (samples prepared under 1 or 5 mbar) decreases, the pore-size increases (up to 6 nm at the end of the CRTA treatment, at 1092°C) and the crystallization into α-alumina takes place. These changes in pore structure give rise to the very special succession of nitrogen adsorption-desorption isotherms presented in Figure 6-8. They correspond to the CRTA treatment of industrial

gibbsite under 0.3 mbar up to the temperatures reported in the Figure and can be interpreted as follows [8]:

- At the very beginning of the thermolysis (183°C) and in spite of a still low BET surface area (ca. 16 m^2g^{-1}) a flat hysteresis loop is visible in the adsorption isotherm, with no final saturation plateau. This was shown to be a characteristic feature of non-rigid, sheet-like, structures [8,12,22]; it therefore shows a slight desagglomeration of the gibbsite which are clearly made of platy crystals visible by electron microscopy.

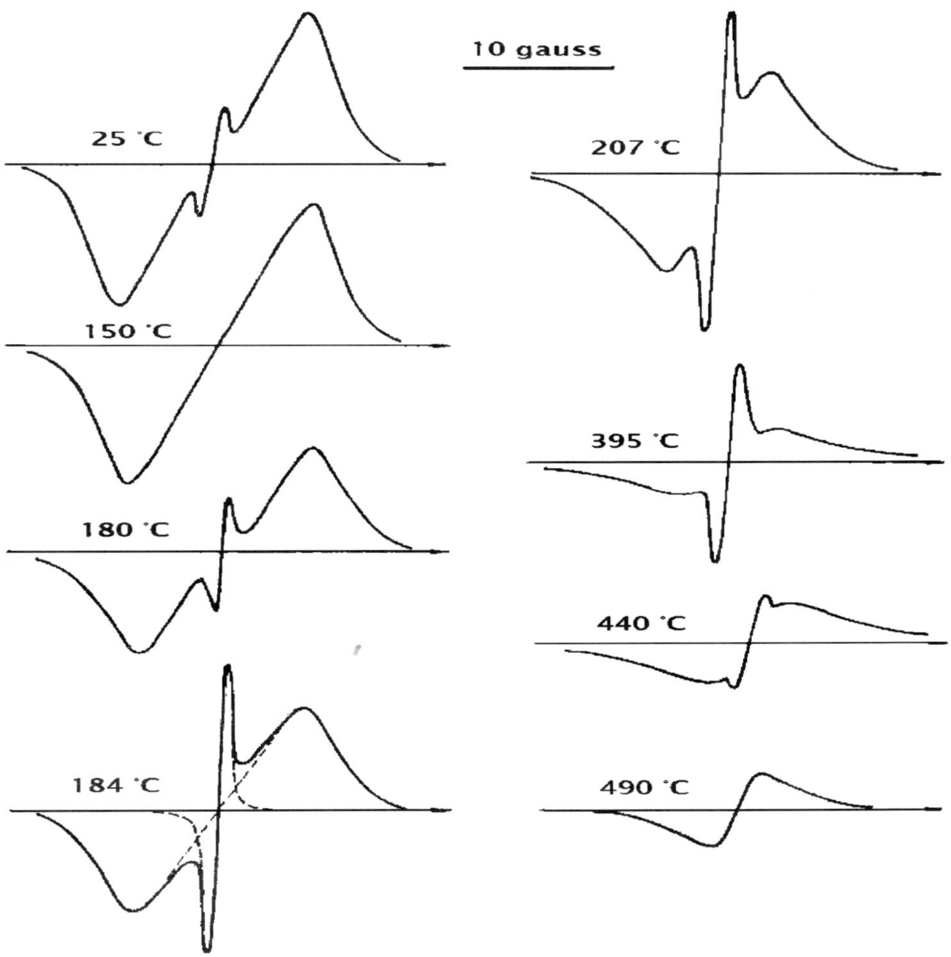

Figure 6-6. Broad band NMR traces recorded at successive stages of the CRTA heat treatment of an industrial gibbsite $Al(OH)_3$ (after [9]).

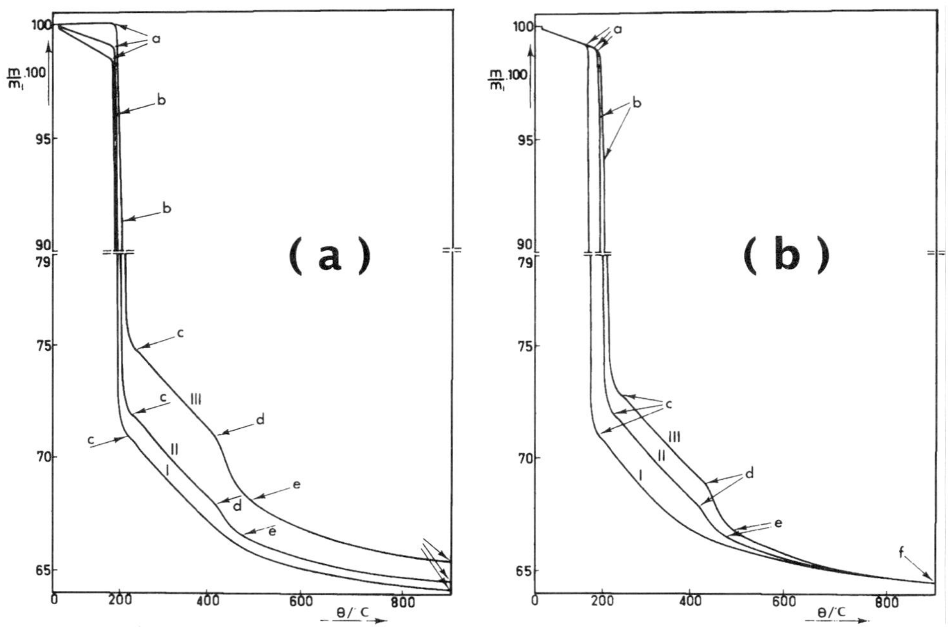

Figure 6-7. CR-EGA curves showing the influence of grain size (0.2, 1 and 80 μm, for curves I, II and III, respectively, obtained under 1 mbar, in Figure (a)) and of water vapour pressure (2.10^{-2}, 1 and 5 mbar, for curves I, II and III, respectively, obtained with the 1 μm sample, in Figure (b)) on the formation of boehmite.

- At the end of what we called step 2) (207°C) a large adsorption takes place at low a pressure, clearly indicating the formation of micropores, as was confirmed by use of the αS method [15,18].
- From there onwards, the action of the further heat treatments is to shift the first "knee" of the adsorption isotherm (i.e. the point where the filling of these micropores by nitrogen is nearly completed) towards higher pressures, indicating a broadening of these pores.
- Finally, for the two last samples (898 and 1092°C) this "knee" enters the hysteresis loop, showing that the micropores have grown enough to become mesopores (i.e. 2 nm width or more). This gives rise to a very peculiar and rare shape of complex hysteresis loop which is the signature of the homogeneity of the CRTA treatment and of the very narrow pore size distribution of the pores which allows them to be filled simultaneously.

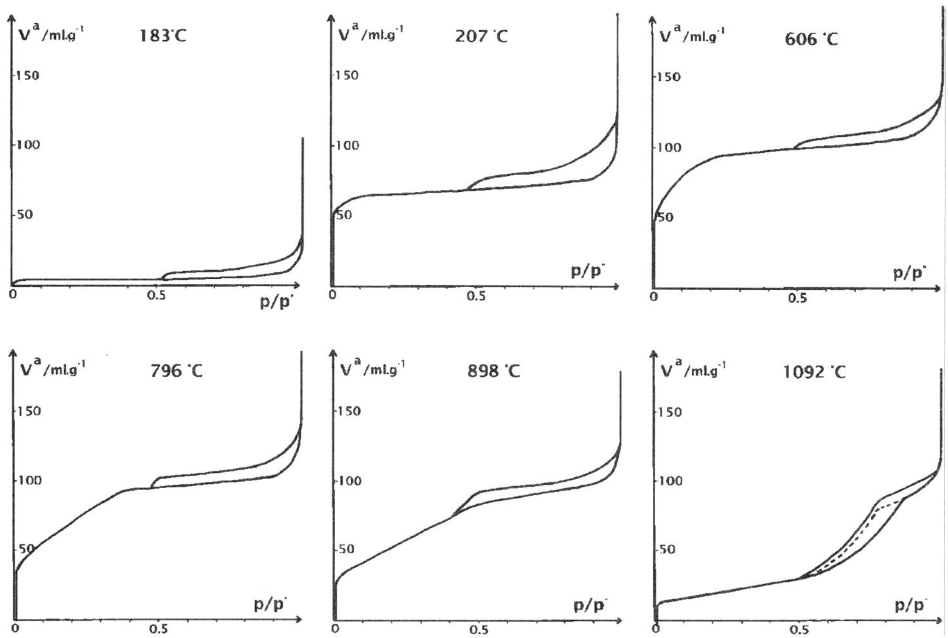

Figure 6-8. N_2 adsorption-desorption isotherms, at 77 K, for an industrial gibbsite sample treated by CR-EGA under 0.3 mbar up to the temperatures reported on the Figure (from [8]).

In the eighties and nineties, a few more studies were carried out by CRTA, in other laboratories, and could bring a useful complement.

Paulik et al. [19] carried indeed the CRTA study of industrial gibbsite thermolysis but, now, under atmospheric pressure and in the presence of air. This work was performed with help of their quasi-isothermal thermogravimetry. To get an idea of the influence of water pressure, three types of crucible were used: the labyrinth-shaped crucible to provide the highest partial pressure of water, a multi-tray crucible to provide the lowest one and, in-between, a bucket-shaped crucible, with or without cover. Although these experiments were more qualitative than the previous ones, since the actual partial pressures of water were unknown, their results fit nicely with those reported above: (i) they still show the major importance of water vapour in the formation of boehmite, with the same pressure dependence (similar increase of the amount of boehmite formed as the vapour pressure increases) and (ii) they still show that the direct transformation of gibbsite into porous alumina (what we have called step 2) is made practically isothermal thanks to the use of a CRTA procedure and that the higher the partial pressure of water, the higher the temperature, for a given rate of decomposition.

Stacey [20] aimed to complement the work by Rouquerol et al. by extending the partial water pressure up to 26.6 mbar, in order to see where the formation of considerable quantities of boehmite is to be expected. He also carried out a complementary kinetic study. His experiments were made with the Constant Rate-Evolved Gas Detection equipment (CR-EGD) with helium gas flow, under atmospheric pressure, with two detectors (catharometer and hygrometer) which he devised and built for that purpose. Under 26.6 mbar he found a boehmite yield of 56% (to be compared with the 30% and 18% yields found by Rouquerol, with a comparable gibbsite, under 1 and 0.08 mbar, respectively [7]). Let us point out, incidentally, one interest of CRTA: although the sample masses used by Stacey ranged between 7 and 27 g, it is still meaningful to compare his results with those from previous experiments carried out with much smaller samples (like the 0.5 to 1 g samples most often studied by Rouquerol et al.) simply because his rates of decomposition were chosen low enough to ensure reasonably low gradients within the sample: his experiments lasted indeed between 14 and 222 hours. Stacey also compared CCl_4 and N_2 adsorption experiments, taking into account the different van der Waals diameter (0.59 nm for CCl_4 and 0.36 nm for N_2) and found that the volume accessible to CCl_4 was 80% smaller than that accessible to N_2, which confirms the existence of the micropores previously detected by the *t*-plot method [20] or the α_S-plot method [13,18]. His nitrogen adsorption-desorption isotherms on samples treated up to 350°C nicely fit between those obtained by Rouquerol at 207°C and 606° (cf Figure 6-8). He interpretes the hysteresis loop in the same way as was done before, i.e. by the existence of slit-shaped mesopores [7,12], but he assumes that the micropores also have the same shape and that the absence of any well-defined maximum uptake shows that some macroporosity is always present. Nevertheless, the authors of this chapter have several reservations about the above interpretation of the adsorption isotherms:

- They find it important to appreciate that this shape of hysteresis loop is typical of *non-rigid* sheet-like structures [7,12,22] (here, resulting from the partial desaggregation of the gibbsite grains on heating); the possible existence of rigid mesopores and macropores ought therefore to be derived from other experiments.
- Since they have clearly shown, in their previous study of the development and shrinkage of the porosity, that the micropores and the hysteresis loop have a different origin and story (cf. Figure 6-8) they don't think that these isotherms bring any support to the idea that the micropores could be slit-shaped. They tend, instead, to consider (i) that

the regular array of dots which is visible in the TEM picture published by Stacey is perfectly consistent with the mechanism of pore formation based on the existence of structural channels regularly located in the gibbsite crystal, (ii) that there is no structural reason to form slit-shaped pores through the basal plane of the crystal (which is pictured here) and (iii) that, as can be seen on the two last adsorption isotherms of Figure 6-8, the complex hysteresis loop thus obtained shows that the mesopores resulting from the broadening of the micropores are clearly different from the non-rigid ones which were giving rise to the initial flat hysteresis; there is therefore no reason for them to have the same shape.

Stacey extended his work to kinetics and applied the rate-jump method to determine the activation energy of gibbsite and boehmite thermolysis. It is most interesting and noteworthy that he found the same value of (272 ± 12) $kJmol^{-1}$ for the two processes. He finally proposed an experimental rate equation for the decomposition of gibbsite into porous alumina and another one for the decomposition of boehmite, which both account for the influence of water vapour pressure.

6.2.2. ACTIVATED CARBON

Activated carbons are probably, with clays, the first adsorbents used by man. Indeed, the charring of carbonaceous precursors (wood, olive stones, cellulose, polymers, peat...) in an oxygen poor atmosphere is at first glance a relatively simple process. However, the complexity and heterogeneous nature of the precursor leads to an even more disordered product. Thus, a number of treatments have been developed to modify the porosity and surface chemistry of the product formed.

The one step process initially used, is now, most often, split into two. An initial step is used to form a carbon skeleton with a fairly low porosity. *Pyrolysis* in an inert atmosphere up to a temperature of 600 to 800°C can be used. Alternatively, this pyrolysis step can be replaced by chemical treatment to render a carbon rich product. This initial step can be followed by a second *activation* step in which an etching of the carbon surface occurs known as burn-off. An increasing burn-off generally leads to an increase in the pore volume and an enlargement of the pore size as well as the pore size distribution. This activation step is often carried out in the range 800 to 1000°C in an atmosphere such as steam or carbon dioxide. A temperature plateau often follows the linear heating. Such atmospheres are used as the reactions that occur are endothermic allowing a degree of control with respect to the exothermic reactions in play when in an oxygen atmosphere.

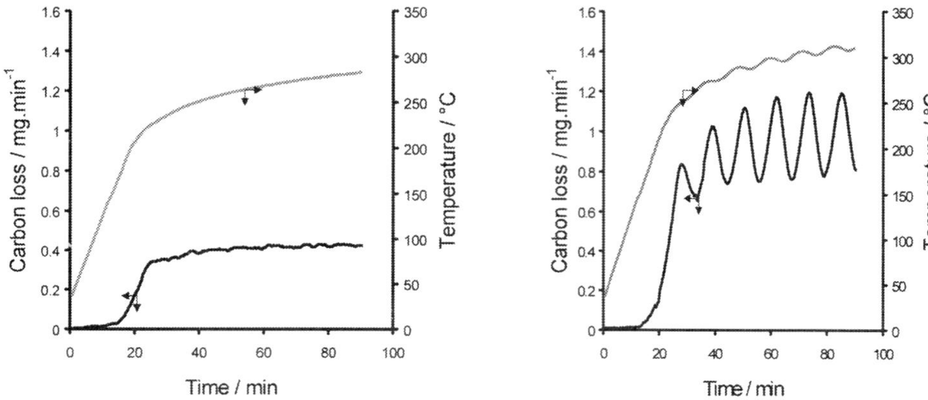

Figure 6-9. Burn off-rate (target = 0.4 mg/min left, 1.05 mg/min right) and corresponding temperature used in Constant Rate Thermal Analysis experiments for the activation of carbon (adapted from [23]).

CRTA can be used to control the two thermal processes of pyrolysis and activation of carbons [23,24]. Controlling the burn-off thus leads to an increased reproducibility. Figure 6-2 shows the temperature and carbon burn-off rates used in a Constant Rate Thermal Analysis experiment at target rates of 0.4 and 1.05 mg/min in an air atmosphere.

Once the target burn-off is reached, the curves in Figure 6-9 show a slow rise in temperature rather than an isothermal plateau. This would suggest a heterogeneity of active sites which react progressively with increasing temperature. One would simply have to stop the reaction at different desired points and determine the surface area and porosity to have a complete picture of the process.

The activation of carbon using the temperature control offered by CRTA seems quite satisfactory at low burn-off rates. However, with the experimental set-up used, at higher burn-off rates the reaction seems more difficult to control due to a probable thermal runaway [23]. The oscillations observed (Figure 6-9) would seem to be a simple problem of thermal lag arising from the heat capacity of the system which could be overcome by a modification of the experimental set-up. However, one may expect ignition of the system at very high rates.

This problem with a simple CRTA set-up lead to a second approach to be developed. It was appreciated that the atmosphere above the sample has an effect on the reaction. The suggestion was therefore to control the amount of oxygen supplied to the system at constant temperature [23]. This gas blending approach follows the carbon dioxide production peak measured in the outlet which is controlled at a constant rate *via* the variation in the inlet, of air

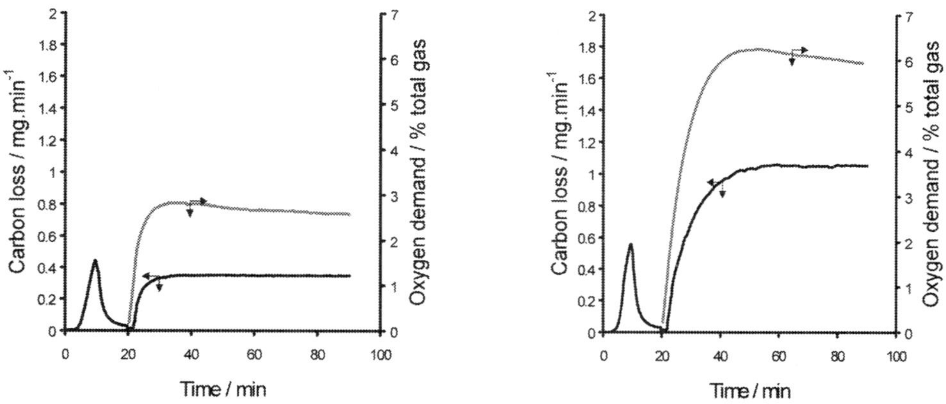

Figure 6-10. Carbon burn-off rates and oxygen demand in constant rate gas blending experiments for target burn-off of 0.35 and 1.05 mg/min (adapted from [23]).

concentration in a nitrogen atmosphere. Figure 6-10 shows the burn-off rate and oxygen demand for target rates of 0.35 and 1.05 mg/min.

It can clearly be seen that this gas blending approach gives a greater control at the higher burn-off rate of 1.05 mg/min with respect to the CRTA control. Indeed, it was demonstrated that gas blending control is effective for burn-off rates above 1 mg/min whereas the temperature control is only effective for rates below 0.4 mg/min.

Furthermore the temperature ranges in which either process is carried out (200–400°C) are much lower than those used traditionally (800–1000°C) leading to a considerable gain in energy. It can be appreciated, as pointed out in Chapter 3, that this approach which is still Sample Controlled is, strictly speaking, out of the scope of thermal analysis, since it is isothermal. Now, since it was inspired by the CRTA "philosophy" and since it complements or replaces a strict CRTA approach, it seemed reasonable to present it here.

6.2.3. ZEOLITES

Zeolite molecular sieves are important catalysts and adsorbents due to the possibility to fine tune their surface chemistry and narrow pore size distribution. They are furthermore doted with high thermal, hydrothermal and acid stability. Their hydrothermal synthesis is usually carried out in the presence of an organic template, which has to be removed to render the porous network accessible. The most common procedure of organic template removal is via thermal degradation (calcination).

This calcination step can lead to the extraction of heteroatoms (Al or P) or the formation of cracks within the crystals (cf. Figure 6-1). Cracks may give rise, not only to parasite adsorption phenomena, but also to a bypassing of compounds during membrane based separation processes. The defects that occur during the thermal extraction of the template result from the considerable build up of pressure gradients within the pores. This arises from the blockage of degradation species formed that are unable to leave due to congestion at the entrances. The most common calcination protocols use linear heating ramps (around 1 $Kmin^{-1}$) up to a final temperature plateau that can be maintained for several hours. This can be carried out under nitrogen, oxygen or mixed N_2/O_2 atmospheres.

The drawback of such protocols is that it, by no means, takes into account the actual degradation reactions within the pores. This can lead to an overlap of different reactions, side reactions as well as a large increase in local pressure gradients within the pores. It would thus seem interesting to use a thermal calcination treatment that can be adapted to the sample, on-line, to the rate of reaction advancement under investigation. With these reasons in mind, several studies have used CRTA methods for the calcination of zeolites and zeolite related materials such as aluminophosphates.

In the case of such materials, degradation reactions involving amines, used as templates, are exothermic. Thus the risk of thermal runaway during calcination is real. The degradation of the tetrapropylammonium template occluded within MFI type zeolites is an example. Figure 6-11 shows the SCTA curves obtained for such a reaction. It can clearly be seen that temperature minima are required to control this reaction.

Figure 6-11 shows the influence of various parameters on the form of the CRTA curve obtained during the calcination of different zeolites and related materials. Indeed, as would be expected, different zeolites, prepared with different templates, give rise to CRTA curves (Figure 6-11a) corresponding to the degradation reactions that occur. In many cases, tertiary or quaternary amines are used as templates which often decompose via Hofmann degradation type reactions.

Even in the cases where zeolites are prepared with the same structure type and with the same organic template, a variation in SCTA curve can be obtained. Silicalite is the pure silica form of the MFI-type structure, however, the silicon atom can be substituted for other heteroatoms such as aluminium (Al-MFI) or iron (Fe-MFI). The SCTA curves (Figure 6-11b) obtained with the aluminium or iron substituted samples differ from that obtained with the pure silica form. It would seem that a second series of reactions occur in the region of the heteroatoms highlighting a catalytic effect of such atoms. Indeed, further studies show [25], as would be expected, the quantity of heteroatoms affects these

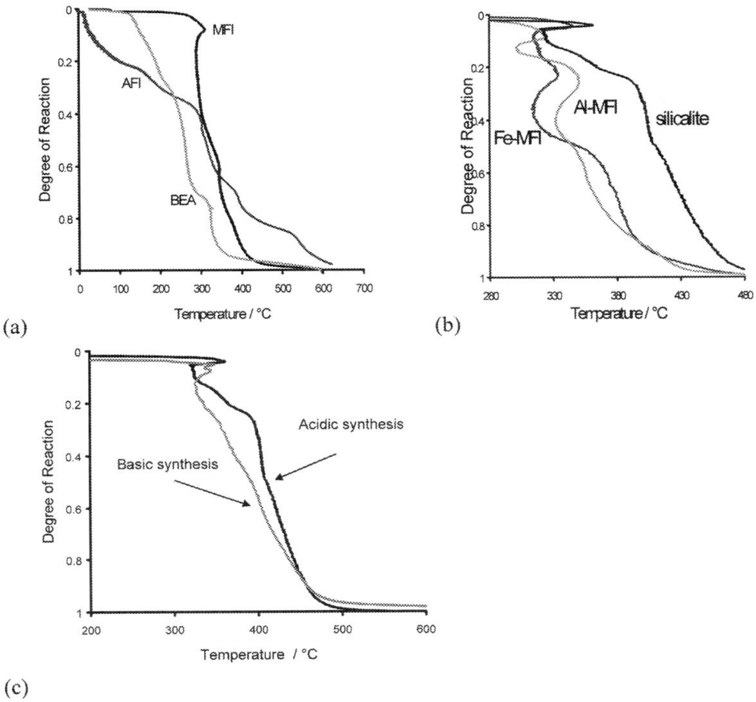

Figure 6-11. SCTA curves obtained for the calcination of various zeolites: (a) various structures (MFI, AFI, BEA), (b) various frameworks, (c) different synthesis media for silicalite.

results. This highlights that the nature and quantity of heteroatom present has an effect on the CRTA curve. Furthermore, the synthesis medium affects the zeolite synthesis, this again is reflected in the SCTA results (Figure 6-11c).

It is interesting to note in the case of the MFI-type zeolites, that several distinct minima in temperature are observed in the CRTA curves. This requirement for the system to cool down so that the reaction can occur at a relatively constant rate suggests an autocatalytic phenomenon for part of the degradation process. If the system is unable to cool down, thermal runaway occurs leading to a build up of pressure within the micropores and eventually to the cracking phenomena observed in Figure 6-1. Thus the use of CRTA here is important in the cases where such defects can be a problem as for example, zeolite membranes.

An advantage in using SCTA is in the understanding of the degradation mechanisms of the organic species present. Indeed, SCTA methods allow the possibility to control the thermal treatment such as to separate various reaction steps that occur within a small temperature domain. The reproducibility attained allows the possibility to prepare samples with the "same thermal history" [26] up to various intermediate points in the transformation. It is thus possible either to analyse *in-situ* the gaseous products, or *ex-situ*, the state of the sample at different points of the transformation. The knowledge of the reactions and side reactions can thus be transposed, allowing modification of more traditional heating protocols.

An interesting example of the use of in situ measurement of the species evolved during the thermal treatment, under vacuum, is shown in Figure 6-12. A mass analyser follows the species evolved during the thermal treatment of dipropylamine template occluded inside $AlPO_4$-11.

The region 2 (Figure 6-12b) from 140 to 210°C is of interest as the mass spectra observed are identical to those obtained with pure dipropylamine under the same experimental conditions. It would seem that the dipropylamine is liberated from the micropores of $AlPO_4$-11 without any degradation.

This may be due to a temperature induced evaporation type mechanism. The dipropylamine molecules are relatively linear and small enough to logically be able to diffuse through the pores. However, as noted earlier, this is not possible under vacuum alone. This temperature region (140–210°C) is higher than the boiling point of the pure liquid (109–110°C under atmospheric conditions) and this is certainly due to the confinement of these molecules, that is to say, due to the interactions between the molecules and the pore walls. Nevertheless, no degradation of these molecules is observed either on leaving the pores or whilst still confined inside.

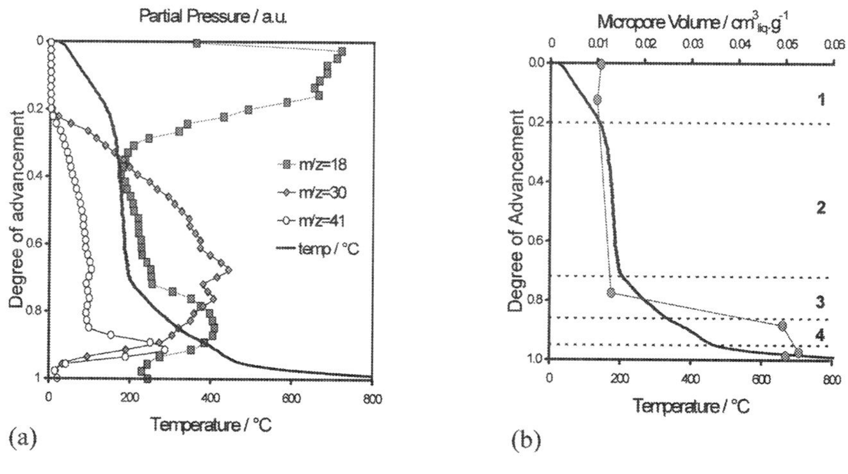

Figure 6-12. SCTA curves obtained for the calcination of AlPO4-11 prepared with dipropylamine: (a) in situ measurement of several species evolved from the sample, (b) ex situ measurement of the accessible micropore volume from nitrogen physisorption (adapted from [27]).

The fact that an evaporation-type mechanism seems to be observed is rather an interesting point. Previous studies of the calcination of zeolites and aluminophosphates have all shown a degradation of the organic template via a Hofmann degradation type mechanism. These solids all have tertiary or quarternary amines as template molecules. Nevertheless, evaporation type mechanisms have been observed for the loss of alkylamines from α-zirconium phosphate intercalates [28]. It would seem that in the present case, as both the aluminophosphate sample has a globally neutral framework and that the secondary amine used here is a far more linear molecule, it is possible for the amine to leave the micropores without degradation even though these pores are relatively small (0.39×0.63 nm^2) with respect to other zeolites studied.

A practical interest of this finding is that, one can envisage the possibility, using CRTA under reduced pressure, of extracting most of the template without degradation before recovery in a cold trap for future use.

In summary, the thermal decomposition using Controlled transformation Rate Thermal Analysis of template is *unique signature for each template-zeolite pair* studied. The shape of this characteristic CRTA curve depends not only on the chemical nature of the T-atom but also to the extent of the Si/T molar ratio of isomorphous substitution. Hence, a CRTA curve could be used to characterise zeolitic samples. Furthermore, CRTA coupled in situ with mass spectrometry permits the correlation of specific reaction steps of a complex mechanism to domains of the corresponding thermogram as well as allowing the identification of various side reactions. In certain cases, CRTA can also be used for template recovery.

6.2.4. ORDERED MESOPOROUS MATERIALS

The structure of the mesoporous molecular sieve, MCM41, comprises of a uniform extended hexagonal pore arrangement. This structure results in the material possessing a large BET surface area, a high porosity and a narrow pore size distribution that can be controlled by the selection of both the synthesis conditions and, moreover, the carbon chain length of the surfactant used. The pore sizes of these materials have been shown to be adjustable from 1.6 to 10.0 nm.

A wide variety of potential applications for these materials are under current investigation which range from catalysis, host-guest chemistry, adsorption and membrane science. It would therefore seem important to understand the mechanisms of formation of such materials to be able to fine-tune the preparative conditions with regard to the desired application. With comparison to the numerous studies that have investigated the synthesis of these materials, little has been reported as to the mechanism by which the surfactant is removed from the mesophase. However this step can considerably alter the final properties of the material. A number of methods for removing the organic phase exist, such as super critical extraction, solvent extraction and ozone treatment.

However, the most common method employed for the removal of the organic template has been, like for zeolite synthesis, by thermal extraction or calcination. As in the case of other adsorbents, CRTA lends itself well to this calcination.

Controlled transformation Rate Thermal Analysis has been used to investigate the mechanism of pore emptying. CRTA allows a controlled rate of elimination of the fragments from the organic template is obtained. This control permits both the temperature and pressure gradients within the sample to be minimised and allows an increased differentiation between reaction steps. It is also possible to couple the CRTA apparatus to a mass spectrometer allowing the evolved gases to be studied in situ during the thermal decomposition of the organic template, cetyltrimethylammonium bromide (CTABr) within the silica framework. An excellent reproducibility can thus be obtained allowing intermediate samples to be isolated and characterised. Solid state MAS NMR studies of the intermediate materials have been used to give further insight into the breakdown of the surfactant structure as the temperature of mesophase calcination was increased. Variation in the porosity, wall thickness and surface hydrophobicity was investigated using nitrogen adsorption, powder X-ray diffraction (XRD) and immersion calorimetry. By using these complementary techniques a great deal of information can be obtained at the different steps of thermal decomposition of the surfactant.

Figure 6-13 shows results obtained for the elimination of cetyltrimethylammonium bromide occluded in a pure silica form of MCM-41. These results are again an example of the complementary use of in situ and ex situ measurements for the understanding of the decomposition mechanisms. The advantage to use CRTA is in the reproducibility in which it is possible to prepare different samples via the same thermal pathway and with the possibility to separate different reactions.

This study [29] suggested that the surfactant (cetyltrimethylammonium bromide) is present in two forms within the inorganic host: the majority which are relatively loosely bound and a small quantity of more strongly bound species which degrade in different temperature domains and via different reaction pathways.

The use of nitrogen physisorption measurements showed the direct formation of the mesoporosity occurs without prior formation of microporosity. From the form of the isotherms, it was suggested that the surfactant residues may form at the channel entrances which act as valve-like structures which *retain the nitrogen* on desorption as unusual hysteresis phenomena were observed. This

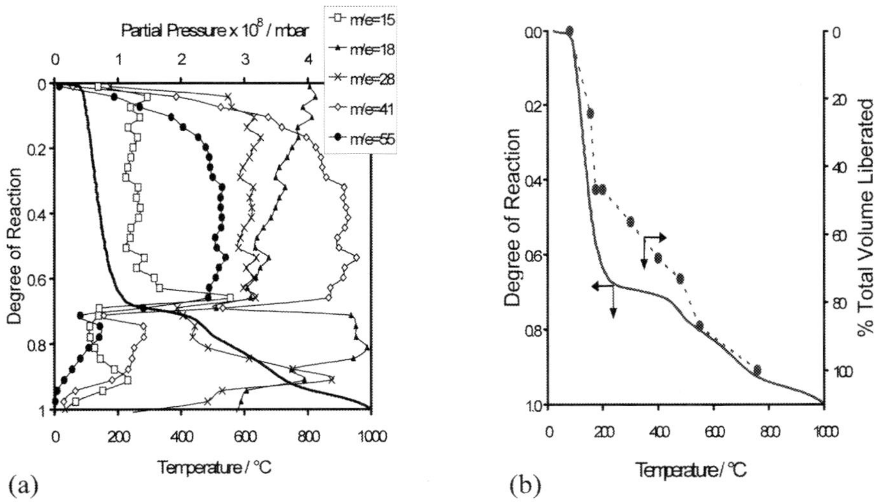

Figure 6-13. Calcination of MCM-41 prepared with cetyltrimethylammonium bromide: (a) evolution of several species followed in situ, (b) evolution of the total pore volume measured *ex situ* by nitrogen physisorption (adapted from [29]).

translates in the pore blocking effect visible in Figure 6-6b which only ends after treatment to 550°C.

Once again, the nature of the surfactant – oxide system synthesised has an effect of the results obtained. In the case of the pure alumina form of MCM-41 prepared with sodium dodecylsulfate [30], part of the surfactant head group is occluded in the walls of the oxide. Thus whilst the degradation of the surfactant chains occurs, a significant porosity is maintained. Treatment to higher temperatures leads to the loss of the head groups which is accompanied by a collapse of the oxide structure. Nevertheless, the use of CRTA allowed the preparation of meterial with both a higher surface area and greater pore volume that materials prepared via traditional calcination protocols [30].

In summary, Sample Controlled Thermal Analysis, most often in the form of Controlled transformation Rate Thermal Analysis can be used in several respects concerning the preparation of adsorbents. Indeed, the above mentioned paragraphs highlight the direct use of sample controlled methods in the preparation of adsorbents. The advantages of such an approach lie in the careful control of the reaction, minimising pressure and temperature gradients as well as avoiding secondary reactions, to form a more homogeneous product. This is shown in the case of zeolite calcination where the number of cracks produced is minimised. A selection of the experimental conditions, notably the pressure above the sample, can lead to adsorbents of varying properties, as in the case of activated alumina. The ability to let not only the reaction dictate the temperature but also the blend of gas, as in the case of carbon activation, leads not only to a control of this process but also a substantial gain in energy.

Whilst it has been possible to prepare up to 200 grams of product using such an approach, production on an industrial scale is limited. However, the use of SCTA in understanding the process and hence to define a heating profile with alternating rapid and isothermal temperature steps has been used. The direct industrial use of SCTA is a goal yet to be achieved.

6.3. SCTA and Adsorbents Characterization

6.3.1. INTRODUCTION

The characterisation of adsorbents can be carried out using Sample Controlled methods. Indeed, in the same way that it is possible to characterise catalysts via the desorption of chemisorbed species, it is possible to characterise adsorbents via the thermodesorption of physisorbed molecules. In such cases, one can glean information about surface area and pore size.

Two main approaches have been taken to characterise adsorbents using SCTA. On one hand, the adsorbent, saturated with liquid, is equilibrated with its saturated vapour. Thus, on heating, a curve is obtained which relates the loss of liquid with respect to the saturated vapour pressures on a flat liquid surface and the liquid inside the pores [31]. This approach has been taken by several authors who have used a Sample Controlled Thermogravimetric approach, namely the Paulik's Derivatograph and is currently known as Quasi-isothermal thermodesorption. Whilst a direct measurement of the weight loss is obtained, the estimation of the actual water vapour pressures is not straightforward. Another approach that is based on Controlled Rate Evolved Gas Detection in which the pressure above the sample (directly related the rate of desorption) is continuously measured and kept constant with time. Whilst the pressure is kept constant the saturated vapour pressure varies with temperature [32]. In each case, the thermodesorption curves are interpreted via the Kelvin equation to give pore size distributions. A further description of the two approaches and selected results is given in the following sections.

6.3.2. QUASI-ISOTHERMAL THERMODESORPTION

The quasi-isothermal heating mode proposed by the Paulik brothers' (and described in Chapter 3), which is used here for the thermodesorption of liquids is one way to increase the resolution of traditional TPD experiments [33]. The Derivatograph was used in all of the quasi-isothermal studies presented in this particular section. The quasi-isothermal heating mode starts here with a linear temperature rise (e.g. 3 K/min). The differential mass loss is calculated with time and when this increases above a certain pre-set value (e.g. 0.5 mg/min), an isothermal plateau is maintained. When the differential mass loss with time descends below this pre-set value, a linear increase in temperature again takes over. The weight loss is thus recorded as a function of temperature.

In thermodesorption experiments, it is important that the vapour pressure above the sample be known. Using the derivatograph, the partial water vapour pressure above the sample is not directly measured. In such cases however, a special "labyrinth" sample cell was developed as schematised in Figure 6-14 in which the cell including sample and excess liquid is covered with an overfitting lid. Such a cell insures that a self-generated atmosphere is maintained above the sample during the experiment and is therefore equal to the atmospheric pressure [33]. During the experiment an increase in temperature leads to an evaporation/boiling of the excess liquid which expels the excess air. The overfitting lid and labyrinth sample cell ensures a relatively long diffusion path of vapour from the sample and avoids any retrodiffusion of air.

Thus during the experiment, it is expected that the water vapour pressure "p" is maintained constant at the atmospheric pressure and a variation in saturation vapour pressure "p°" occurs with temperature "T". Schematically, a curve is obtained of the form shown in Figure 6-15. The first half of the curve is due to the loss of the excess liquid and has nothing to do with the sample. It can be used to calibrate the instrument temperature. The following part of the curve is due to the emptying of the various classes of pores. An increase in temperature leads to the progressive emptying of pores of ever decreasing widths.

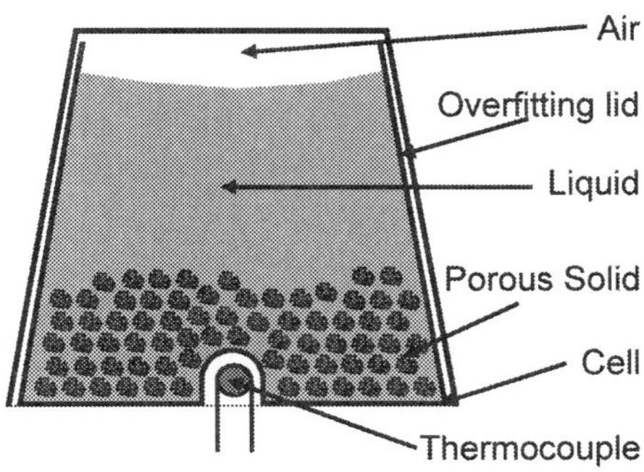

Figure 6-14. Schematic representation of the form of crucible used in quasi-isothemal thermogravimetric experiments.

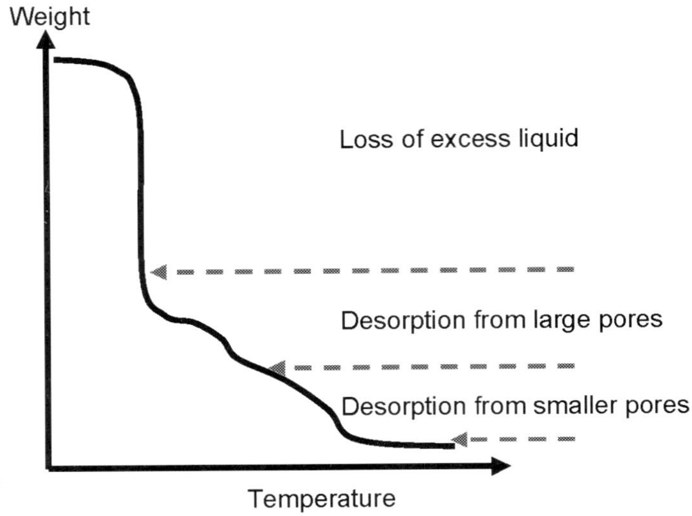

Figure 6-15. Schematic representation of a curve obtained from a quasi-isothermal gravimetric experiment with a solid with a bimodal pore size distribution.

An analogy can thus be made between the results obtained via this method and those obtained from physisorption measurements. This is highlighted below in Figure 6-16 in which the co-ordinates of the two graphs are adjusted such that the forms of the curves are similar.

One can thus appreciate how it is possible to calculate the pore volume and even the pore size distribution in analogy with adsorption calculations. Thus, the weight loss observed for each step can be related to the pore volume and the temperature range in which the weight loss occurs can be related to the pore width via the Kelvin equation. Supposing perfect wetting, the equation takes the form:

$$r = \frac{2\gamma V_M}{RT \ln\left(p/p^\circ\right)}$$

Here, in addition to p, p° and T mentioned above, one has to take into account the surface tension of the liquid " γ ", the molar volume "V_m", both of which also vary with temperature. The radius "r" obtained via this expression is a *core radius* and one has to additionally take into account a surface layer which is not desorbed.

This layer is dependent on the type of surface, the size of pore and on the preadsorbed species used. Indeed, as in adsorption, the thickness of the surface layer increases with pore size. In several cases, the relationship put forward for

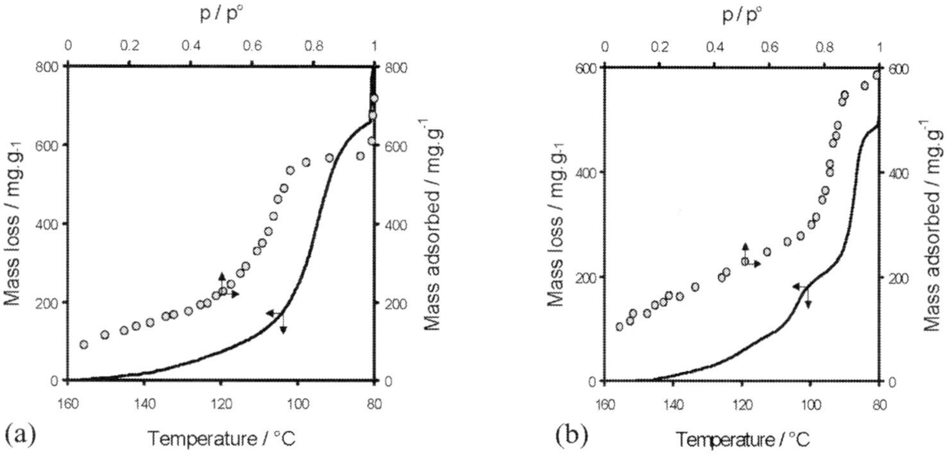

Figure 6-16. Comparison of desorption curves of benzene (full line) obtained from quasi-isothermal thermogravimetry, and nitrogen (points) obtained via physisorption: (a) silica gel, (b) mixture of two silica gels of different pore sizes (adapted from [34]).

adsorption experiments can be used in thermodesorption. This is the case, for example, for water-silica system [35]. In other cases, the difference in pore size distribution compared with that obtained for nitrogen has been used [36].

The surface under investigation can lead to differences in the thickness of the adsorbed layer. In the case in which grafted solids are under investigation, one can also observe the case where the preadsorbed species would seem to migrate between the chains of the organic grafts leading to discrepancies. This was shown to be the case for benzene on a C18 chain grafted silica [36].

In a number of cases, good agreement has been obtained in pore volume and pore size distribution between quasi-isothermal gravimetry and nitrogen physisorption. The following Table 6-1 regroups a number of systems that have been studied by this method.

The choice of liquid used for the characterisation via thermodesorption is important. The liquid has to wet the solid sufficiently and one has to avoid specific interactions with the surface. A number of different wetting liquids have been used as can be seen in the table above. Most of the studies gave satisfactory results suggesting good wetting. Nevertheless, it has been noted that in the case of silica, water is unsuitable as it forms a stable surface film [39]. However in other studies, water has been found to be quite satisfactory [31, 38].

The SCTA curves obtained via quasi-isothermal gravimetry can be treated to yield desorption energy profiles. This theoretical treatment is detailed in reference [31] for the case of heterogeneous surfaces.

Table 6-1. Several systems studies using quasi-isothermal thermogravimetry

Type of sold	Vapour	Reference
Silica	Benzene	[36], [37], [38]
	n-butanol	[31], [34]
	n-hexane	[39]
	CCl4	[40]
	Water	[31], [38], [39]
Alkyl grafted silica	Methanol	[40]
	Benzene	[36]
	n-butanol	[36]
Alumina	Water	[38]
	Benzene	[38]
Clay intercalates	hydrazine hydrate	[41]
Active carbon	benzene	[38], [42]
	Methanol	[42]
	Water	[31], [38]
	n-propanol	[40]
Porous glass	Benzene	[40]
Porous Polymer	CCl4	[43]
	Water	[44]
Polymer foil	benzene, acetone	[45]

One advantage of using a saturated system for thermodesorption is the possibility to characterise the sample in the wet state in the actual liquid used in the application itself. A second advantage can be for the characterisation case of fragile solids. This can be the case for organic films [45] and porous polymers [43], which shrink on vacuum and are brittle at low temperature making nitrogen physisorption experiments unmeaningful. The results obtained with quasi-isothermal gravimetry thus can be compared with those obtained using thermoporometry.

Whilst in most cases the quasi-isothermal method is quite a satisfactory method for the characterisation of porous solids, it would seem that some discrepancies are observed with respect to other methods such as nitrogen physisorption or mercury porosimetry. In some cases, this may be due to phenomena such as adsorbent swelling or migration between organic chains as mentioned above. However, it would seem that this method has difficulty in detecting small amounts of porosity [40] which may simply be due to the experimental conditions chosen which may lead to a too rapid desorption of wetting liquid. Thus the sample has difficulty in creating its own atmosphere. Furthermore, it would seem that difficulties also arise in the case of the characterisation of micropore volumes [42]. This has been explained by the difficulty in the observation of the end of pore emptying. This may be due to a difference in the pressure above the sample with respect to the assumed saturation vapour pressure. This variation in pressure above the sample at the end of the experiment may be due to the special form of sample cell used and difficulty in the actual measurement of the vapour pressure above the sample.

6.3.3. CONSTANT RATE THERMODESORPTION

Although quasi-isothermal gravimetry measurements lead to an increased resolution with respect to standard TPD, several limitations can be observed as suggested above. For small amounts of a given porosity, the quasi-isothermal programme is not at its optimum [40]. Thus for the evaluation of microporosity and adsorbed surface species, the same resolution is obtained as with a linear heating programme [40]. Although a special "labyrinth" sample cell was used, the partial pressure is not measured and more importantly, not controlled. Finally, the experiments always require an excess of liquid to be used. This allows pore volumes and pore size distributions to be calculated but estimations of surface area are not possible.

The above mentioned points suggest that some improvements can be made with respect to quasi-isothermal thermogravimetric measurements.

One possibility is to use evolved gas analysis instead of thermogravimetry. This can be done using the apparatus developed by Rouquerol [46]. For such experiments, the sample is firstly outgassed via SCTA under vacuum to clean the surface and pores of impurities. The cell is then attached to an apparatus for the preadsorption. The sample and pure liquid are placed at different temperatures to induce the relative pressure above the sample required. Thus for calculation of the surface area, a relative pressure corresponding to monolayer coverage is used. For information on the total pore volume and pore size distribution, a relative pressure around 0.95 is required.

After preadsorption, the sample cell is isolated and transferred to the SCTA apparatus adapted for evolved gas analysis [46]. Such experiments have been carried out under constant reduced pressures in the range from 10^{-3} to 5 mbar. This is in contrast to the quasi-isothermal thermogravimetric experiments carried out under atmospheric pressure. Here, once again, the pressure "p" is maintained constant and the saturated vapour pressure "p°" varies with temperature "T". The Kelvin equation can be used to calculate pore size distributions as described above. One therefore relates the desorption *temperature* to the pore size via the Kelvin equation, instead of relating the pore size to the desorption *pressure*, as is usually done in the isothermal Barrett, Joyner and Halenda method [22].

The other experimental conditions are adjusted to give a constant weight loss of around 0.14 to 0.5 mg/h which is again different from that used in the quasi-isothermal thermogravimetric measurements (30 mg/min). This rate of weight loss is calculated in a separate experiment. This much lower rate leads to an increased resolution allowing small amounts of porosity to be investigated. This resolution has been required in one study where it was required to follow the small amount of porosity created during the surface corrosion of metallic alloys [47]. An interesting point is that the fluid used to characterise this porosity is the same as that involved in the ageing process.

Such experiments have been used for the study of thermodesorption of water from various zeolites [48] as well as from heterogeneous surfaces [49,50]. Here a residual pressure above the sample of the order of 10^{-2} mbar was used which is sufficient to estimate the micropore volume as well as to highlight the influence of cation sites on the thermodesorption curves [48]. Even though the temperature at which the experiments were started was around –30°C, this is not sufficient to retain physisorbed water within mesopores. Two solutions are possible: the first is to drastically increase the residual pressure above the sample as above; the second is to decrease the initial temperature at which the experiments start.

Whilst the former solution is very well exploited by the quasi-isothermal method, the latter solution has been adopted for experiments carried out under reduced pressure. A schematic diagram of the apparatus is shown in Figure 6-17 [32]. The furnace used for the experiments is placed onto a liquid nitrogen Dewar. Inside the Dewar, an electrical resistance maintains a constant boiling of the liquid nitrogen. The vapour thus produced passes through a spiral tube within the furnace assuring constant cooling down to a minimum temperature of 163 K. The furnace resistance is connected to a PID regulation whilst the sample temperature is measured *via* a 100 Ω platinum probe. This apparatus permits a temperature interval from 163 to 473 K to be explored. The maximum time taken for an experiment depends on the amount of liquid nitrogen in the dewar. The current set-up, with the 25 L reservoir allows experiments to be carried out for up to 20 h.

Several results for the thermodesorption of water are given in Figure 6-18.

In the case of the mesoporous sample MCM41, the water desorption curve is shown in Figure 6-18a. The initial temperature of –80°C was low enough under the conditions of residual pressure (10^{-3} mbar) to avoid vapour loss prior to heating. This curve can be treated as described in reference [32], in an analogous manner to that described above for quasi-isothermal measurements, to give the pore size distribution in Figure 6-18b (dark circles). This compares well with the pore size distribution obtained from the BJH treatment of the desorption branch of the physisorption isotherm at 25°C (light circles). These results not only show the use of SCTA for the characterisation of porosity under

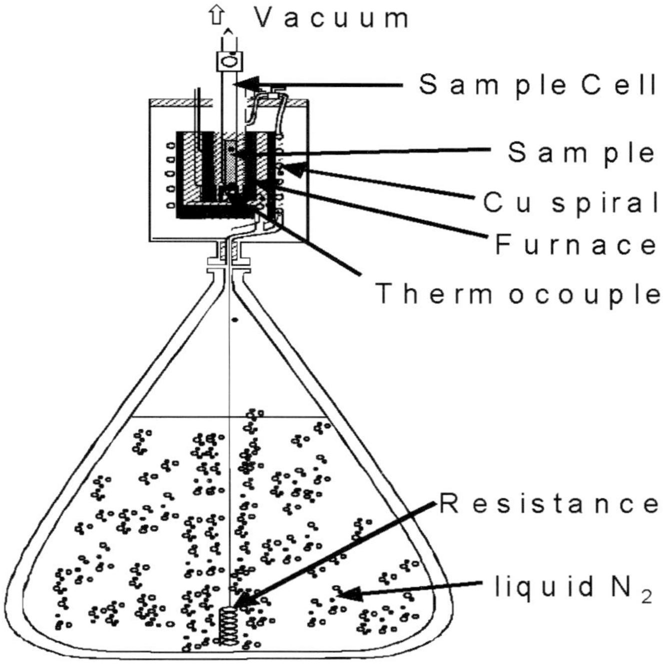

Figure 6-17. Schematic diagram of the apparatus developed for thermodesorption studies starting at low temperature [32].

isobaric conditions but also that any preferential interactions that water would have with the silica surface of MCM-41 has no effect on the result obtained here.

In the case of the mesoporous sample MCM41, the water desorption curve is shown in Figure 6-18a. The initial temperature of –80°C was low enough under the conditions of residual pressure (10^{-3} mbar) to avoid vapour loss prior to heating. This curve can be treated as described in reference [32], in an analogous manner to that described above for quasi-isothermal measurements, to give the pore size distribution in Figure 6-18b (dark circles). This compares well with the pore size distribution obtained from the BJH treatment of the desorption branch of the physisorption isotherm at 25°C (light circles).

The SCTA curve shown in Figure 6-18c is obtained for the thermodesorption of water from the aluminophosphate $AlPO_4$-5. Logically, this microporous sample (pore diameter 0.73 nm) desorbs water at a higher temperature than MCM-41.

Nevertheless, this desorption occurs well below room temperature highlighting the interest in using the set-up in Figure 6-16. Figure 6-18d shows the Kelvin radius that is obtained directly from the corresponding SCTA curve.

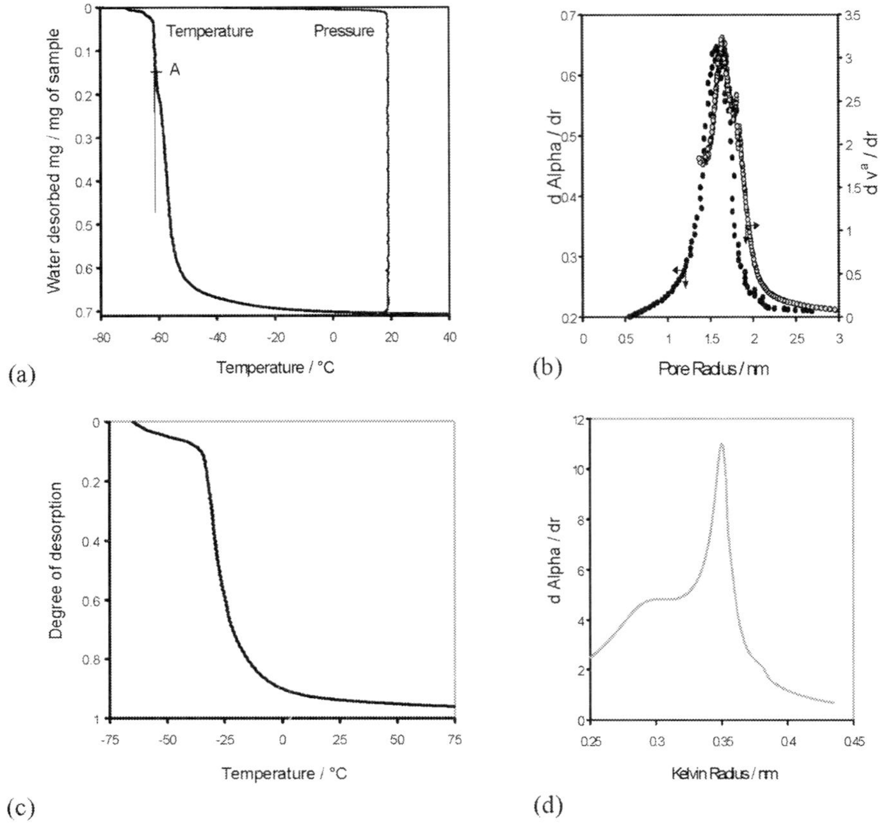

Figure 6-18. Constant Rate Evolved Gas Detection (CR-EGD) curves obtained for the thermodesorption of water from MCM-41 (a) and from AlPO4-5 (c). (b) pore size distributions for MCM-41 from SCTA (dark circles) and from water desorption at 25°C (light circles). (d) Kelvin radius obtained for $AlPO_4$-5 from the treatment of the CR-EGD curve (c).

It is surprising that the value obtained centred on 0.35 nm is not far from the theoretical value of 0.36 nm. However, both the use of the Kelvin equation and neglecting an eventual surface adsorbed layer renders such a result purely of interest.

It is also possible to preadsorb at a partial pressure in which a monolayer is adsorbed, before desorbing under SCTA conditions to calculate the amount desorbed and thus the specific surface area. This can be of interest when the surface area under investigation is low or the amount of sample that can be studied limited. This can be the case where a thin, porous oxide layer is formed during the corrosion of metals. In such cases, the resolution of 2 cm^2 that can be attained using this approach is of great interest [32,47].

The thermodesorption of other liquids than water can be used in the case of zeolites to characterise the chemical nature of the surface active sites. One suggestion has been the use of isopropylamine which would seem to desorb in two regimes from HY and NaY zeolites [51]. Here, the use of the rate jump method was used to characterise these two regimes. The first regime corresponds to the physisorbed species whereas the second regime corresponds to the species chemisorbed on Brønsted acid sites. Interestingly, the 20% difference in apparent activation energies obtained from CRTA and linear heating experiments was explained by diffusion effects that occur in the latter measurements. Complementary information on the catalytic characterisation of such materials can be found in Chapter 7.

Controlled rate thermodesorption under reduced pressure has been applied to the characterisation of active carbons [52]. In this case, phenol was used as the desorption fluid. In contrast with traditional linear heating TPD experiments [53,54], the SCTA experiments showed that no cracking of phenol occurred and for the carbon studied, three domains were identified and characterised witht he aid of high resolution argon physisorption experiments.

Finally, it is possible to further exploit the SCTA thermodesorption curves obtained with heterogeneous solids using statistical rate theory of interfacial transport [50]. The theoretical basis is detailed in ref [50] in which two different approaches are given for SCTA experiments. Each approach leads to a different condensation energy distribution, calculated assuming equilibrium or mixed equilibrium/non-equilibrium conditions. An example of water desorption from hydroxapatite is given in which physisorbed and chemisorbed species are highlighted.

6.4. SCTA and Adsorbent Outgassing

The aim of adsorbent outgassing is to prepare a well-defined, reproducible and meaningful surface on which the adsorption experiment will be carried out. This is different from, and more difficult to define than a "perfect cleaning" of the surface, which is rarely obtained on a technological adsorbent without bringing irreversible changes (of the structure, texture or surface chemistry), so that the sample studied is not any more meaningful from the view point of its envisaged application. Also, the outgassed adsorbent must be ready to withstand, without undergoing any further change, the vacuum usually requested at the beginning of the adsorption experiment. This is why vacuum outgassing tends to be the most used. Now, the questions one should answer prior to carrying out any outgassing in a conventional gas adsorption bulb, are the following:

- How could we avoid the "bumping" or spurting of fine powders (especially when micron-sized) towards the vacuum line at the temperatures when the rate of outgassing is high? The "practical answer" which is to use a frit or a wool pad to retain the powder is of course fully inconsistent with a vacuum treatment since experience shows that pressure can then easily overpass several (and even, sometimes, 10 or 20) mbar…
- How should we select the right outgassing temperature?
- Should we end or not by an isothermal plateau?
- Is it a means to get a thermal analysis curve exactly corresponding to the outgassing?

The use of Controlled Rate Evolved Gas Detection (CR-EGD) actually brings an answer to all questions above, since:

- The spurting of the powders on outgassing is simply fully avoided through the permanent control of the rate of gas evolution. A complementary set-up, which we describe elsewhere [22] can even avoid this problem during the initial room-temperature evacuation.
- There is no need of any frit, so that the residual pressure (or "vacuum") in the close surroundings of the sample is comparable to that measured, above, by the gauge.
- The equivalent of a TG curve is permanently recorded during the CR-EGD treatment.
- It is wise to check, on this curve, that the frequent outgassing temperature of 140–150°C (which is usually enough to eliminate the physisorbed species, yet insufficient to alter the adsorbent) (i) does not correspond to any modification of the sample and (ii) corresponds to a minimum slope of the CR-EGD curve, so that minor experimental errors on the outgassing temperature should not affect the state and mass of the sample.
- In case a different outgassing temperature is to be chosen, this should be done, ideally around an inflexion point of the CR-EGD curve (showing that the evolution of physisorbed species is completed and that the modification of the sample has not yet started).
- The nearly "quasi-equilibrium" state of the sample during the CR-EGD outgassing at a low rate makes that, when the final temperature is reached, there is no need for an isothermal plateau.

Moreover, as illustrated in Figure 6-19, this allows to keep on the CR-EGA curve proper the point representing the state of the sample. It can be said that samples 3, 4 and 5 (obtained by CR-EGD) directly result from each other, whereas samples 1 and 2 (obtained by a conventional heat treatment, with a final plateau), have no direct filiation.

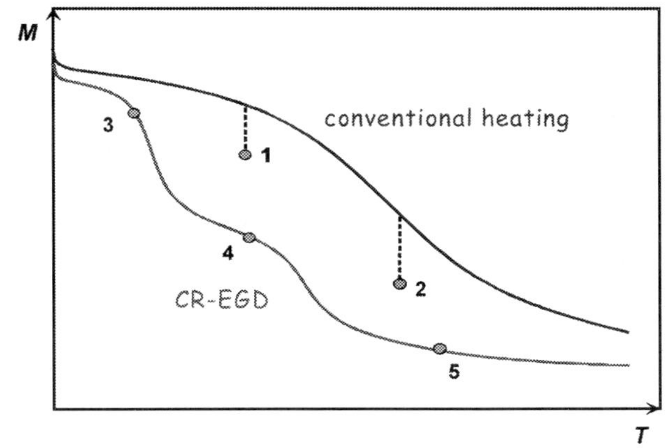

Figure 6-19. The conventional heat treatment (upper curve) ends by an isothermal plateau, up to points 1 or 2. The CR-EGD treatment (lower curve) allows to directly stop the outgassing at points 3, 4 or 5 (after [22]).

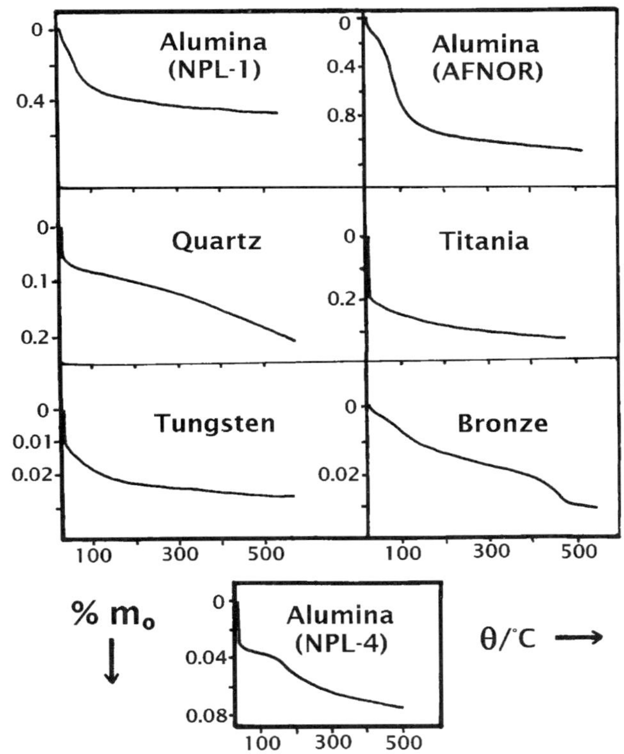

Figure 6-20. Vacuum CR-EGD curves for 7 low-surface area samples (after [55]).

An application of the above guidelines is illustrated in Figure 6-20 in the case of a number of potential reference materials for low surface area (0.1 to 10 m^2 g^{-1}), which were examined in the scope of a BCR programme [55]. The curves reported where recorded during CR-EGD experiments carried out under a constant pressure of 3.10^{-5} mbar, with a controlled outgassing rate of 0.1 mg lost *per hour.* For most samples, at the exception of the two last ones (bronze, *ca* 0.1 m^2g^{-1}, and alumina NPL-4, *ca* 0.12 m^2g^1), a final outgassing temperature of 150°C is satisfactory. Now for the bronze, a clear step is visible around 450°C, in spite of the very small loss experienced by this sample (only 0.03% mass loss between 25 and 525°C, which illustrates, in passing, the high sensitivity and resolution of this type of CR-EGD). The explanation was brought by XRD, which showed that an annealing was taking place at 450°C, accompanied with the disappearance of one of the two cubical face centred phases observed in the lower temperature range. In this case, an outgassing up to 500°C provides a more stable state of the sample.

For the alumina-NPL 4 sample, the situation is more critical, since the step observed is around 150°C, i.e. exactly at the temperature normally chosen for outgassing. Here, again, XRD showed that this step corresponded, in spite of the moderate temperature, to a recrystallisation of the sample (initially containing 3 corindon-type crystalline phases and one amorphous phase). For such a sample the outgassing temperature finally selected could be safely taken as 100°C [56].

6.5. Conclusions

The examples just given illustrate how SCTA (mostly, actually, in the form of Controlled Rate Thermal Analysis) is well-suited to control the thermal preparation and outgassing of adsorbents and to characterize them. It proves indeed to be a powerful tool:

- to obtain homogeneous conditions throughout the sample
- to see the influence of the controlled parameters (specially the residual or partial pressure of the surrounding gases)
- to understand the mechanism of the transformation (usually with help of a kinetic study) thanks to the high synchronism of the transformation in all grains of the sample
- to derive the best conditions to obtain a tailor-made porous solid
- to derive the approximate conditions to be fulfilled in an industrial furnace (temperature profile, partial pressure) to obtain the desired adsorbent.

It can be said that, ideally, the industrial process itself should be sample-controlled. Now, this is usually difficult to achieve because of the long response time of the system, which makes it relatively difficult to obtain a stable regime of the heating control. In reality, the major interest of SCTA is when you don't know yet the sample and when the experiment automatically provides you with the temperature profile corresponding to the transformation rate selected to achieve the desired homogeneity. Once the behaviour of the sample is known, then it is enough, at the industrial scale, to mimic the temperature and atmosphere conditions of the genuine SCTA experiment. Conversely, it is also possible, once the SCTA study is completed, to model the transformation in any point of an industrial furnace by taking into account, for example, the actual pressure and temperature gradients, specially if they can be measured by sensors appropriately located.

References

1. C.S. Smith, Trans. *A.I.M.E. (Metal Division)* 137 (1940) 23.
2. H. Palmour, D.R. Johnson, in "*Sintering and Related Phenomena*", G.C. Kuczynski, N.A. Hooton and C.F. Gibbon (Eds.), Gordon and Breach, New-York (1967) 779.
3. H.C. Stumpf, A.S. Russell, J.W. Newsome and C.H. Tucker, *Ind. Eng. Chem.* 42 (1950) 1398–1403.
4. S.J. Gregg and K.S.W. Sing, *J. Phys. Chem.* 55 (1951) 592 & 597.
5. J.H. de Boer, J.M.H. Fortuin and J.J. Steggerda, *Kkl. Nederl. Akad. Wetensch. Proc., B.*, 57 (1954) 434.
6. D. Papée and R. Tertian, *Bull. Soc. Chim. Fr.,* (1955) 983.
7. J. Rouquerol, Thesis, Faculty of Sciences of Paris University, 19th November 1964 (*Série A, No. 4348, No. d'ordre 5199*).
8. F. Rouquerol, Thesis, Faculty of Sciences of Paris University, 20th December 1965 *(Série A, No. 4654, No. d'ordre 5501).*
9. J. Rouquerol, J. Fraissard, J. Elston and B. Imelik, *J. Chim. Phys.*, 4 (1966) 607.
10. J. Mayet, J. Rouquerol, J. Fraissard and B. Imelik, *Bull. Soc. Chim. Fr.*, (1966), 2805–11.
11. J. Rouquerol, *J. Therm. Anal.,* 2 (1970) 123.
12. F. Rouquerol, J. Rouquerol and B. Imelik, *Bull. Soc. Chim. Fr.*, 10 (1970) 3816.
13. M. Ganteaume and J. Rouquerol, *J. Therm. Anal.,* 3 (1971) 413–20.
14. J. Rouquerol, *J. Thermal Analysis*, 5 (1973) 203–16A.
15. M. Ganteaume, Thesis, Université de Provence, Marseille, 24th February 1973.
16. J. Rouquerol, F. Rouquerol and M. Ganteaume, *J. Catal.,* 36 (1975) 99–110A.
17. J. Rouquerol and M. Ganteaume, *J. Therm. Anal.*, 11 (1977) 201–210A.
18. J. Rouquerol, F. Rouquerol and M. Ganteaume, *J. Catal.,* 57 (1979) 222–30A.
19. F. Paulik, J. Paulik, R. Naumann, R. Köhnke and D. Petzold, *Thermochim. Acta,* 64 (1983).

20. M.H. Stacey, *Langmuir,* 3 (1987) 681.
21. B.C. Lippens and J.H. de Boer, *J. Catal.,* 4 (1965) 319.
22. F. Rouquerol, J. Rouquerol and K.S.W. Sing, "Adsorption by powders and porous solids: principles, methodology and applications", Academic Press, London, New-York (1999) 467 pages.
23. E.A. Dawson, G.M.B. Parkes, P.A. Barnes, M.J. Chinn and P.R. Norman, *Thermochim. Acta*, 355 (1999) 141–146.
24. J. Díaz-Terán, P. Llewellyn, J. Rouquerol, A.J. López-Peinado and A. Jerez, Proceedings of the 6th Meeting of the Spanish Carbon Group, 23–25 oct. 2001, Cáceres-Mérida, Spain.
25. C. Sauerland, P.L. Llewellyn, Y. Grillet and F. Rouquerol, In: *Proceedings of the 12th Int. Zeolite Conference*, (M.M.J. Treacy, B.K. Marcus, M.E. Bisher & J.B. Higgins (Eds.), MRS, Warrendale (USA), 1999, pp. 1707–1714.
26. J. Rouquerol, *Thermochim. Acta*, 144 (1989) 209.
27. N. Dufau, L. Luciani, F. Rouquerol and P. Llewellyn, *J. Mat. Chem.*, 11 (4) (2001) 1300–1304.
28. K. Peters, R. Carleer, J. Mullens and E.F. Vansant, *Microporous Materials*, 4 (1995) 475.
29. M.T.J. Keene, R. Gougeon, R. Denoyel, P.L. Llewellyn, R.K. Harris and J. Rouquerol, *J. Mat. Chem.*, 9 (1999) 2843–2850.
30. L. Sicard, P.L. Llewellyn, J. Patarin and F. Kolenda, *Micro. Meso. Mat.*, 44–45 (2001) 195–201.
31. V.I. Bogillo and P. Staszczuk, *J. Therm. Anal. Cal.*, 55 (1999) 493–510.
32. V. Chevrot, P.L. Llewellyn, F. Rouquerol, J. Godlewski and J. Rouquerol, *Thermochim. Acta*, 360 (2000) 77–83.
33. J. Goworek, W. Stefaniak and A. Dąbrowski, *Thermochim. Acta*, 259 (1995) 87–94.
34. J. Goworek and W. Stefaniak, *Colloids Surfaces A*, 134 (1998) 343–347.
35. H. Naono and M. Hakuman, *J. Coll. Interf. Sci.,* 145 (1991) 405.
36. J. Goworek and W. Stefaniak, *Colloids Surfaces A*, 80 (1993) 251–256.
37. J. Goworek, W. Stefaniak and A. Dąbrowski, *Thermochim. Acta*, 259 (1995) 87–94.
38. P. Staszczuk, *J. Therm. Anal.*, 53 (1998) 597–616.
39. J. Goworek, W. Stefaniak and M. Prudaczuk, *Thermochim. Acta*, 379 (2001) 117–121.
40. J. Goworek and W. Stefaniak, *Thermochim. Acta*, 286 (1996) 199–207.
41. J. Kristóf, R.L. Frost, W.N. Martens and E. Horwáth, *Langmuir*, 18 (2002) 1244–1249.
42. J. Goworek and W. Stefaniak, in "*Thermal Analysis of Active Carbons*", p. 39–46.
43. Z. Hubicki, J. Goworek and W. Stefaniak, *Bull. Pol. Acad. Sci. Chem.*, 42 (2) (1997) 169–176.
44. J. Goworek, W. Stefaniak and W. Zgrajka, *Mat. Chem. Phys.*, 59 (1999) 149–153.
45. J. Goworek and W. Stefaniak, *Colloids Surfaces A*, 82 (1994) 71–75.

46. J. Rouquerol, S. Bordère and F. Rouquerol, in "*Thermal Analysis in the Geosciences*", W. Smykatz-Kloss, S. St. J. Warne (Eds.), Springer-Verlag, Berlin, 1991, pp. 134–150.
47. J. Godlewski, A. Giordano, V. Chevrot, P. Llewellyn, F. Rouquerol and J. Rouquerol, CEA Technical Note No. DEC/SECA/LCG/98.003, June 1998.
48. M.J. Torralvo, Y. Grillet, F. Rouquerol and J. Rouquerol, *J. Therm. Anal.*, 41 (1994) 1529.
49. P.A. Barnes, G.M.B. Parkes, D.R. Brown and E.L. Charsley, *Thermochim. Acta*, 269/270 (1995) 665.
50. F. Villiéras, L.J. Michot, G. Gérard, J.M. Cases and W. Rudzinski, *J. Therm. Anal. & Cal.,* 55 (1999) 511–530.
51. E.A. Fesenko, P.A. Barnes, G.M.B. Parkes, D.R. Brown and M. Naderi, *J. Phys. Chem. B*, 105 (2001) 6178–6185.
52. L.J. Michot, F. Didier, F. Villiéras and J.M. Cases, *Pol. J. Chem.*, (1997) 665–678.
53. J. Rivera-Utrilla, M.A. Ferro-Garcia, C. Moreno-Castilla, I. Baautista-Toledo and J.P. Joly, *J. Chem. Soc. Farad. Trans*. 91 (1995) 3213.
54. M.A. Ferro-Garcia, J.P. Joly, J. Rivera-Utrilla and C. Moreno-Castilla, *Langmuir*, 11 (1995) 2648.
55. J. Rouquerol, F. Rouquerol, M. Triaca and O. Cerclier, *Thermochim. Acta*, 85 (1985) 305.
56. F. Rouquerol, J. Rouquerol, G. Thevand and M. Triaca, *Surface Science*, 162 (1985) 239.

Chapter 7

SCTA AND CATALYSIS

E.A. FESENKO, P.A. BARNES and G.M.B. PARKES

Centre for Applied Catalysis, University of Huddersfield, UK

7.1. Sample Controlled Thermolysis

Usually catalysts and their precursors are prepared by temperature programmed thermolyses carried out isothermally for a given time at a predetermined temperature, which is reached using a linear heating rate. Under such conditions, uncontrolled temperature and pressure gradients are created in the system and the reaction rates vary significantly during the preparation procedure [1]. Hence, catalysts obtained at the beginning and the end of conventional thermolysis, when the reaction rate is low, are made under very different conditions from those prevailing when the reaction rates are at their highest level. Furthermore, it is well known that rates of many thermal reactions are influenced by the partial pressure of product gases that vary from one instrument to another and lead to irreproducible reaction environments. SCTA techniques can be applied with advantages to avoid these problems and produce catalysts in a reproducible and uniform manner, with pre-determined properties [2].

7.1.1. DECOMPOSITION OF COMPLEX PRECURSORS USING VARIOUS SCTA TECHNIQUES

Often complex catalysts are prepared via the thermal decomposition of an intimate mixture of precursors. In such cases the process is frequently complicated as it involves various reactions that may overlap. Furthermore, the self-generated atmosphere formed during an earlier step may influence later processes. In this way the composition of the product and, hence, its catalytic activity, can be dependent on the preparation procedure [3].

A mixture of nickel, magnesium and calcium hydroxides can be used as a model system because it produces three distinct, but overlapping, peaks and so demonstrates the abilities of the different SCTA approaches to the preparation of catalysts via precursors that involves complex thermal processes [4]. The conventional decomposition of an equal mass mixture of nickel, magnesium and calcium hydroxides, using a linear heating rate of 10 $Kmin^{-1}$ (Fig. 7.1), shows the expected three separate hydroxide decomposition steps and a small initial evolution of water. A key feature of linear heating is the large variation in the reaction rate over each decomposition step from zero, at the start of the process, to a maximum value at the peak. Moreover, the water evolution of the first events generates additional pressure gradients that influence the reaction rate of the consequent processes. Hence, in such cases the properties of the final catalyst may well depend not only on the unknown and uncontrolled temperature and pressure gradients created during the particular decomposition but also on the nature of proceeding reactions.

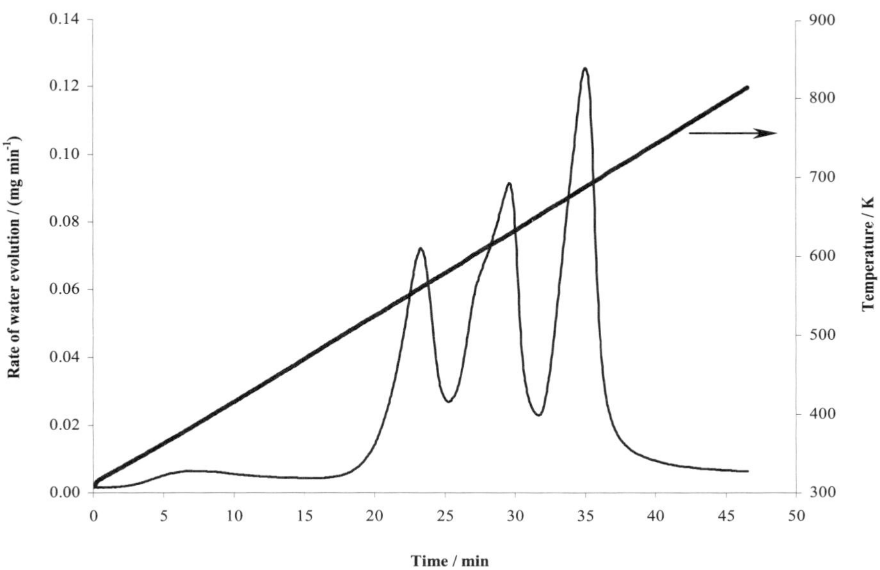

Figure 7.1. The decomposition of 5 mg of an equal mass mixture of nickel, magnesium and calcium hydroxides using a linear rate of 10 $kmin^{-1}$.

The decomposition of the mixture using CRTA (Fig. 7.2) may appear more complicated at the first sight than the corresponding linear heating process, and occurs via several steps. The first, represented by the sharp initial peak, is due to the rapid loss of adsorbed water at low temperatures. This small initial process appears as a transient phenomenon here since it is completed before the selected 'target' reaction rate is achieved. The three plateaux represent the thermal events giving rise to the main peaks observed in the linear heating experiment. They show that the rate of water evolution, and hence the rate of process, is maintained at a constant value for the duration of each transformation, apart from small initial and final deviations that can be avoided by lowering the reaction rate. The deeps are due to the decrease in the level of water vapour between the three reactions, and the drop in the signal at the end of the run is due to completion of the final decomposition. To maintain the reaction rate constant throughout the decomposition processes the temperature during the reactions rises very slowly and, in this case, never quite becomes isothermal [5]. The temperature regimes required to maintain a constant reaction rate are not the same for the three events, reflecting the different kinetics involved in each step. The third decomposition event is characterised by a small initial positive peak on both the evolved water and temperature curves. It could be that the main reaction here is either preceded by a significant nucleation stage, which

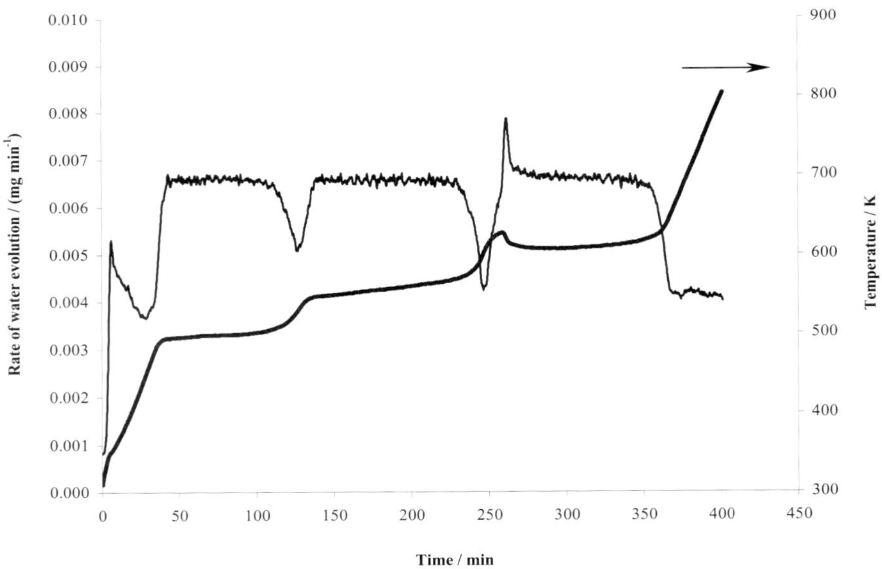

Figure 7.2. The decomposition of 5 mg of an equal mass mixture of nickel, magnesium and calcium hydroxides using the CRTA technique, with a heating rates varying between ±10 Kmin^{-1}.

has a larger apparent activation energy, or that autocatalysis is involved. Apart from providing detailed information on the mechanism of the reactions, the CRTA technique produces catalysts under the very desirable conditions of controlled temperature and pressure gradients, whilst minimising the influence of the self-generated atmosphere on the subsequent processes in a reproducible and uniform manner [6].

In the SIA technique the heating rate does not change continuously as the measured sample response varies but is switched between zero (isothermal) when the sample response exceeds an upper preset threshold, and a fixed value when the response falls below a lower preset threshold [7]. In this case each step of decomposition of the mixture occurs at a constant temperature (Fig. 7.3). Here, the initial water loss, which is too small to trigger, is also followed by the three main reactions. The first two processes produce large peaks of an unusual shape, which differ from that of the third, again reflecting the dissimilarity of the kinetics involved. It should be considered that, as the reaction rate is not constant, there are variable pressure and concentration gradients across the sample. So, in this case, the temperature gradients are reduced, but the pressure gradients are uncontrolled. However, the overlapping consecutive processes have less influence on each other as they are more efficiently separated when compared with a conventional linear heating.

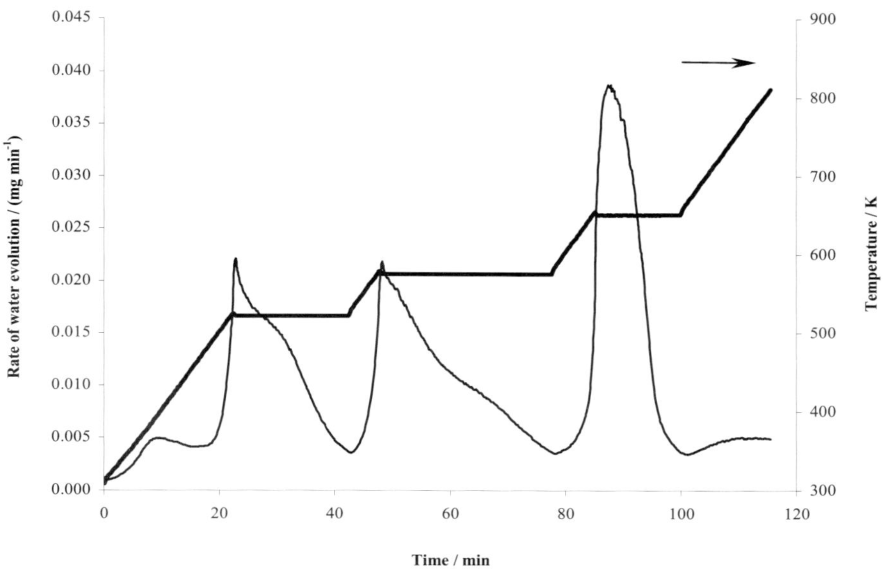

Figure 7.3. The decomposition of 5 mg of an equal mass mixture of nickel, magnesium and calcium hydroxides using the SIA method, with a linear heating regime of 10 Kmin^{-1}.

The CRTA and SIA methods operate by controlling the sample heating as some function of the difference between the measured reaction rate and a preset 'target' reaction rate [8]. A disadvantage of this is that it is difficult to set a single target level which is equally appropriate for complex systems that may produce events of greatly different magnitudes, as explained in Chapter 3. The PSH method eliminates such problems by changing the heating rate at the same relative point on the peak (i.e. the same value of α) irrespective of its absolute magnitude [9]. During the decomposition of the mixture using PSH method (Fig. 7.4), the heating rate is switched to zero when the rate of change of the reaction rates switches from positive to negative, i.e. immediately following the peak maximum. Once the first process reaches completion and the following event is detected, the heating rate returns to its original value. This method, which works in the opposite way to CRTA and SIA, involves heating quickly through thermal events and slowly between them, and so significantly increases the resolution of the two peaks, with respect to time, with little apparent distortion of their respective shapes. It can be seen that the switching point between heating and isothermic conditions is at the same relative point on each peak, despite their difference in size. Relatively high temperature and concentration gradients are created in the system during each step when the

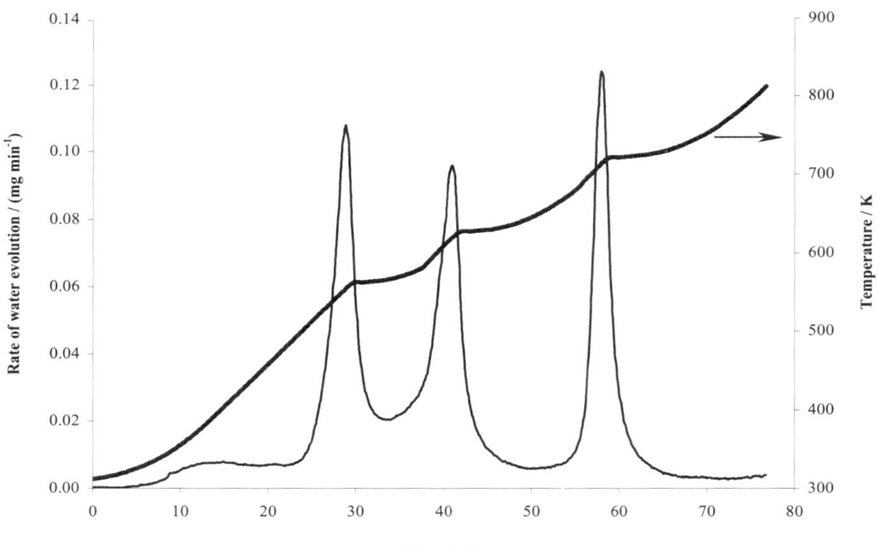

Figure 7.4. The decomposition of 6.2 mg of an equal mass mixture of nickel, magnesium and calcium hydroxides using the PSH technique, with a variable heating regime between 0 and 10 Kmin^{-1}.

PSH method is used. However, they do not influence the subsequent processes unduly as, in this case, each stage is virtually completely separated from the adjacent ones. The individual peaks are slightly narrower (in terms of time) than those for the linear heating experiment, as each peak has less contribution from the other in its profile. Moreover, unlike conventional linear heating thermolysis, the maximum reaction rate of the second peak is smaller than that for the first one, which may be interpreted as being due to the influence of water evolution occurring during the decomposition of nickel hydroxide on that of the subsequent magnesium hydroxide reaction.

The advantages and disadvantages of the above SCTA approaches over conventional linear heating techniques in the study of catalyst preparation are more clearly demonstrated in alpha (α) plots which show the extent of the decomposition as a function of time or temperature [10] (Fig. 7.5). In conventional thermal analysis, where the heating rate is constant, the resolution of the thermal events in either the time or temperature domains is essentially the same. However, this is not so for SCTA methods, where the relationship between time and temperature is always non-linear when a reaction is occurring. Consequently, SCTA α-curves displayed as a function of time look markedly different from those plotted as a function of temperature. The reason for the better resolution of the consecutive events can be seen by examining the temperature traces for the techniques used. While the temperature increases at a fixed and predetermined rate in conventional linear heating experiments, it changes only very slowly in CRTA during the main reaction stages and is isothermal for SIA. The reduced heating rate gives rise to the improved resolution, as the thermal events are resolved into sharp steps when α is plotted against temperature. The beginning and end of the reactions are more clearly delineated when the experiments are carried out using the SCTA techniques. However, sometimes, as in the present example, the two stages in the third decomposition, revealed in the CRTA experiment, are not detected using SIA under conditions employed here. By enforcing a constant temperature for each decomposition event any such detail is lost in the resulting vertical line on the α-temperature profile. It should be noted that selection of other parameters for the SIA experiment could produce a markedly different profile in which each decomposition process are split into several smaller heating and isothermal stages, possibly giving rise to artefacts. One limitation of both CRTA and SIA is the large increase in time taken to complete an experiment compared with that required to complete the equivalent conventional linear heating run using a typical heating rate of 10 $\mathrm{Kmin^{-1}}$. By contrast, the PSH techniques improve resolution in the minimum amount of time. However, PSH α-temperature profiles appear particularly unusual as half of each of the two processes occurs under linear heating while the other half occurs under isothermal conditions.

Nevertheless, as this technique produces enhanced resolution in the time domain, it is particularly valuable when the real-time separation of complex gas evolution mixtures is required for further downstream analysis.

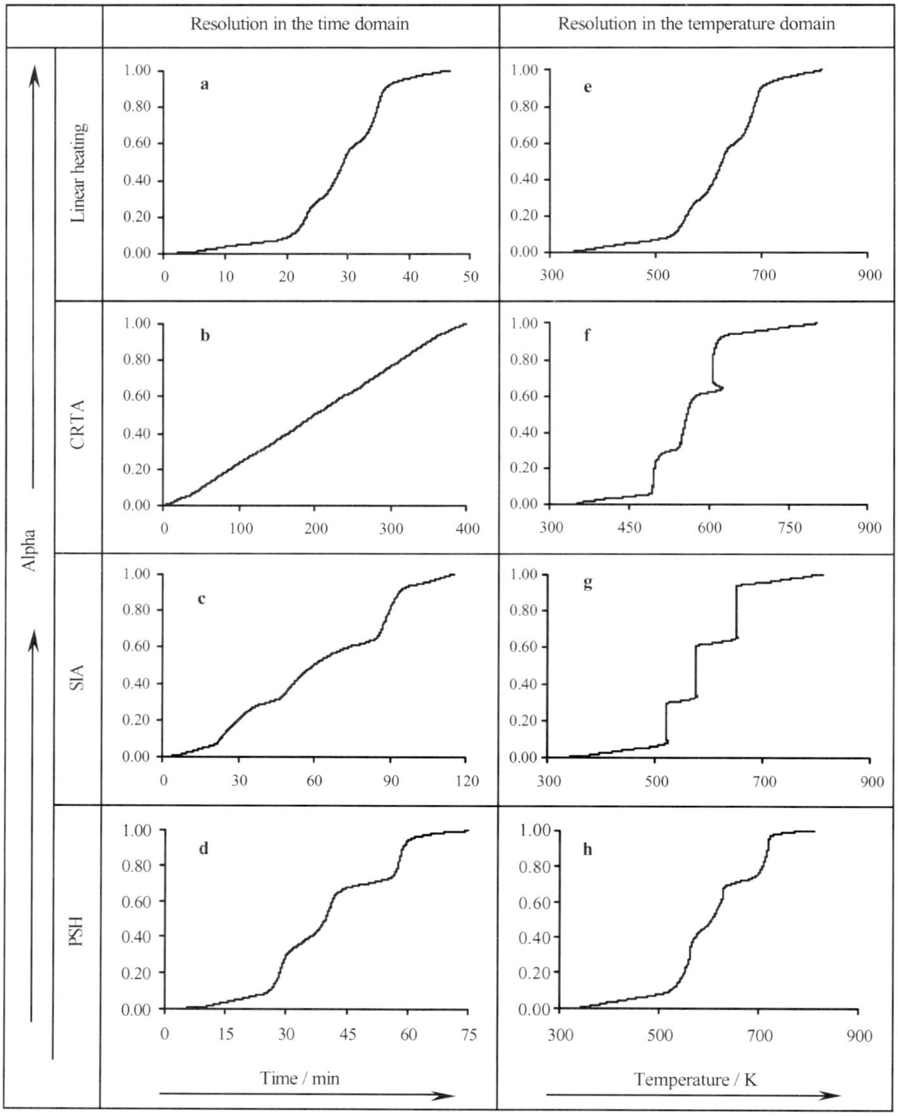

Figure 7.5. Comparison of SCTA methods: alpha plots for the thermal decomposition of the mixture as calculated from the results presented in Figures 7.1–7.4.

The application of various SCTA methodologies to catalysts can provide a precisely controlled environment during their preparation [11]. As some of the these techniques are designed to achieve this aim, the choice of an appropriate SCTA method depends on the particular material and on whether increased resolution is required in the time or the temperature domain throughout transformations of a complex precursor.

7.1.2. VARIATION IN REACTION PATHWAY DURING CRTA PREPARATION OF CATALYSTS

Whatever technique is used in thermal preparation, the quality of the catalyst is determined by its activity and selectivity toward a specific reaction. It is well known that, in many cases, the reactivity of a material is directly related to the surface area of either the catalyst (e.g. Raney nickel) or to that of a specific component (e.g. copper surface area in methanol synthesis catalysts). Such properties of the product depend very markedly on the method of preparation and the CRTA approach provides a invaluable insight into the processes involved. For example, one of the ways of producing reactive, high surface area nickel oxides is via thermolysis of nickel nitrate hexahydrate:

$$Ni(NO_3)_2{\cdot}6H_2O \rightarrow NiO + 2NO_2 + \frac{1}{2}O_2 + 6H_2O$$

In this case, the properties of the final oxide are affected by quantity of the intermediate products developed during the process. Distinguishing a kinetic law for the nickel nitrate transformation to the oxide by means of conventional thermal analysis is difficult and the experimental data found in the literature for this reaction varies considerably [12] The results obtained using CRTA provide evidence for a significant effect of the residual pressure above the sample that explains these differences. The work of Llewellyn et al. [13] on this process establishes that the thermal reaction pathway and intermediate products, along with the texture of the final catalyst, clearly vary with the residual pressure in the system. Hence, the properties of the final product can be optimised, in a reproducible manner, by controlling the pressure and gaseous environment at each point of the thermolysis.

The first conclusion for the thermolysis of nickel nitrate hexahydrate under CRTA conditions is that increasing the residual pressure during the preparation of the oxide raises the temperature of the process as expected. However, changing the pressure in the system leads to very different intermediate products. When the thermolysis proceeds under high pressure, at 10 kPa, the reaction occurs in one step without any suggestion of intermediate products. On the other hand, if the process is carried out at a pressure of 0.5 kPa, the reaction occurs via four steps, whereas at a lower pressure of 5×10^{-3} kPa, five steps can be identified, allowing the separation of five and six thermolysis products respectively. The temperatures at which the intermediates are isolated and also the gases evolved during the each step in the CRTA thermolysis vary with the residual pressures (Table 7.1).

Decreasing the controlled pressure from 0.5 to 5×10^{-3} kPa significantly changes the pathway of the thermolysis. Only water is evolved in the first four steps of the process at pressure of 5×10^{-3} kPa suggesting the dehydration of the hexahydrate to tetrahydrate, dihydrate, monohydrate and anhydrous nickel nitrate, respectively. The fifth step, accompanied by a characteristic evolution of

Table 7.1. Properties of intermediate and final products obtained using CRTA thermolyses of nickel nitrate hexahydrate [13]

Pressure / kPa	Final temperature / K	Gases evolved	Surface area / ($m^2 g^{-1}$)	Chemical composition of the products
5×10^{-3}	293	H_2O	0	Nickel nitrate tetrahydrate
	318	H_2O	9	Nickel nitrate dihydrate
	383	H_2O	16	Nickel nitrate monohydrate*
	468	H_2O	18	Anhydrous nickel nitrate
	483	H_2O, NO	80	Nickel oxide
	723	H_2O, NO	58	Nickel oxide
0.5	308	H_2O	<1	Nickel nitrate tetrahydrate
	343	H_2O	<1	Nickel nitrate dihydrate
	443	H_2O, NO	6	$Ni(NO_3).2Ni(OH)_2$
	513	H_2O, NO	<1	Sub-stoichiometric nickel oxide
	773	H_2O, NO	<1	Stoichiometric nickel oxide
10	853	NA	<1	Nickel oxide

* It is worth noting that nickel nitrate monohydrate was first isolated using CRTA techniques.

NO, indicates the decomposition of the nitrate to the oxide. Hence, lowering the pressure of the thermolysis allows the separation of these two reactions as, under these conditions, the dehydration of nickel nitrate hexahydrate is complete before the denitration starts. Further heating of the product from 483 K up to 723 K leads to both a decrease in the specific surface area and increase in grain size of the catalyst.

The decrease in surface area of the nickel oxide formed at 5×10^{-3} kPa during the further heating can be explained by considering the mechanism of decomposition of anhydrous nickel nitrate in the controlled environment of CRTA [14]. Figure 7.6 shows simulated SCTA alpha curves using the Avrami-Erofeev kinetic model with an exponent n = 2. In both the cases it is taken that the Arhenius preexponential factor and the activation energy are 10^8 min^{-1} and 160 $kJmol^{-1}$, respectively. In the CRTA simulation, a value of 2×10^{-2} min^{-1} refers to the reaction rate, that is maintained constant throughout the process, while in the corresponding SIA this value represents the point when the heating rate, at 10 $Kmin^{-1}$, switches to an isothermal period. Both the models define a characteristic feature, that in the first, case is the minimum temperature of the

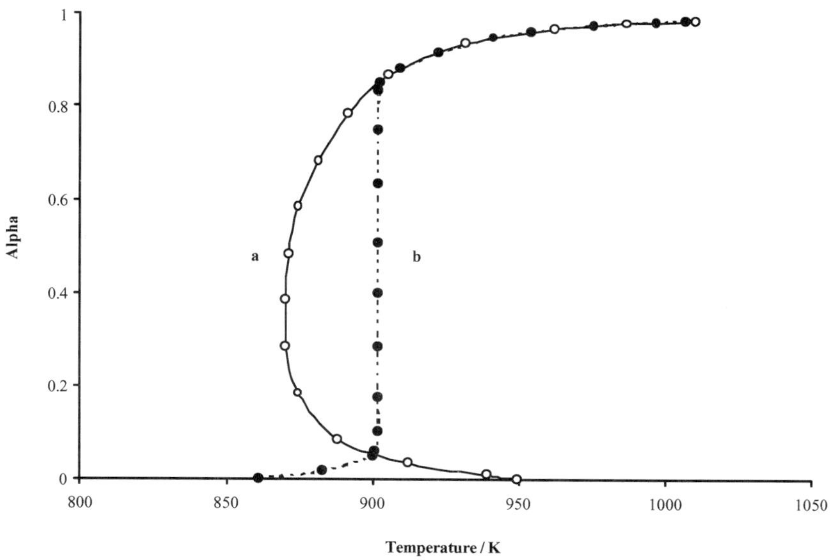

Figure 7.6. CRTA (a) and SIA (b) simulated curves for the thermal decomposition of anhydrous nickel nitrate calculated for an A_2 kinetic model, assuming the following kinetic parameters: $E = 160$ $kjmol^{-1}$, $C = 2\times10^{-2} min^{-1}$, $\beta = 10$ $Kmin^{-1}$ [15].

alpha plot (α_{min}) or attributes to a maximum reaction rate during SIA isothermal period (α_{max}). Moreover, the agreement between these values suggests that the process undergoes the same kinetic mechanism in both the techniques. The experimental results of Gotor et al. [15] on the decomposition of anhydrous nickel nitrate are consistent with a process which obeys the Avrami-Erofeev kinetic law despite the SCTA technique used. This mechanism involves instantaneous nucleation on the surface of the crystal, with subsequent growth of the nuclei into the particles, so additional heating leads to sintering of the final nickel oxide.

The products of the low pressure CRTA thermolysis of nickel nitrate hexahydrate are characterised by remarkably high surface areas (80 m^2g^{-1}) while apparently the same reaction carried out under higher pressures gives products with an undeveloped surface. Such differences in the surface areas of the products formed using the controlled conditions of CRTA provide a clear evidence that reaction pathways can vary according to the residual pressure. The thermolysis of nickel nitrate under a controlled pressure at 0.5 kPa undergoes different reaction sequences to those occurring at lower pressure. They also commence with the dehydration of the nitrate to the tetrahydrate and dihydrate, the water being evolved in just the first two steps. During the next stage NO is detected as well as H_2O and the amount of nitrous species greatly increases in the fourth and final steps of the reaction. Hence, the denitration in this case begins before the dehydration is completed, explaining the formation of $Ni(NO_3)_2.2Ni(OH)_2$ at the end of the third step. The water evolution during the thermolysis of this basic salt explains the simultaneous decomposition of both the hydroxide and nitrous compound in the fourth stage to form a sub-stoichiometric nickel oxide that transforms to a stoichiometric oxide at higher temperatures.

At a residual pressure of 10 kPa the transformation of nickel nitrate hexahydrate to the oxide occurs in a single step, via a solid state fusion of the reactant, apparently without any intermediate products. The final product, formed at highest temperature of 853 K, has negligible surface area. It can be clearly seen that the properties of catalysts obtained using calcination are directly dependent on the exact conditions of their preparation process, when the processes involved are influenced by pressure and temperature gradients arising in the system during the reaction. This shows that nickel oxide with a high surface area can be produced if the dehydration of the precursor is separated from the subsequent decomposition demonstrating significant advantages of the CRTA in producing a detailed insight into the complex processes often found in catalyst preparation.

7.1.3. TEXTURE AND STRUCTURE OF CATALYSTS PREPARED USING CRTA

Porous aluminas were the first catalyst supports whose preparation and mechanism of formation were studied under CRTA conditions: since they are also largely used as adsorbents, their case is treated in Chapter 6.

Porous iron oxides such as hematite (α-Fe_2O_3), magnetite (Fe_3O_4) and maghemite (γ-Fe_2O_3), which are important in various areas, including catalysis, can be synthesised via the following solid state reactions [16]:

$$\alpha\text{-FeOOH} \rightarrow \alpha\text{-Fe}_2\text{O}_3 \rightarrow \text{Fe}_3\text{O}_4 \rightarrow \gamma\text{-Fe}_2\text{O}_3$$

In this process, lath-like goethite (α-FeOOH) crystals undergo topotactic dehydroxylation during which the original single crystals of precursor are replaced by a highly ordered aggregate of small α-Fe_2O_3 crystals with a large amount of pores. Subsequent reduction and oxidation are also topotactic, i.e. the products retain the shape and size of their precursors while the texture and structure of the products depend on experimental parameters such as the annealing temperature and atmosphere.

Table 7.2 summarises results of Perez-Maqueda et al. [17] for the decomposition of goethite, with an initial BET surface area of 43 m^2g^{-1}, and shows the increase in this value for the products obtained by linear heating in various environments. These results seem to be contradictory, as there is an opposite effect of the heating rates on the development of surface area of the

Table 7.2. Specific surface areas of hematite prepared by decomposition of goethite using conventional linear heating [17]

Atmosphere	Heating rate / ($K\ min^{-1}$)	Specific surface area / ($m^2\ g^{-1}$)
Vacuum	0.5	83
	20	101
Static air	0.5	101
	20	77
	$\infty^{(a)}$	56
Closed tube	$\infty^{(b)}$	49

(a) The sample was introduced directly into a furnace preheated to 623 K.
(b) The sample (in a closed tube) was introduced directly into a furnace preheated to 623 K, giving an estimated water vapour pressure of ca. 100 kPa.

iron oxides in vacuum and in air. Oxides with high surface areas are produced in vacuum at the high heating rate of 20 K min^{-1}, while to obtain the material with the same porosity using similar thermolysis, but in static air, the precursor needs to be decomposed at the low heating rate of 0.5 K min^{-1}. In conventional thermal analysis an increase in heating rates leads to greater reaction rates and also to increased product vapour pressure in the proximity of the sample. In vacuum, where the pumping minimises the influence of water vapour pressure, increasing the reaction rate has a crucial effect on developing the porosity of the products. However, if the self-generated gas is not removed from the reaction interface, increasing the reaction rate raises the water pressure in the system which has an adverse effect on the specific surface areas of the iron oxides. In such cases, where the reaction rates and partial pressure of the evolved gases have an opposite effect, it is difficult, if not impossible, to predict the texture and structure of the final products without precise control of both parameters. However, creating a tailored porosity in such catalysts is possible by applying CRTA methods to the preparation procedure, as both the reaction rate and partial pressure of product gas are controlled.

Table 7.3 shows the properties of hematite samples prepared under various CRTA conditions in the thermolysis of goethite [18]. As is shown in Section 7.1.2, varying the reaction environment can influence reaction pathways and the transformation of precursor to an iron catalyst is also affected by the conditions under which the decomposition occurs. At a low controlled pressure

Table 7.3. Specific surface area of hematite prepared by decomposition of goethite using CRTA techniques [18]

Controlled pressure / kPa	Reaction rate / (min^{-1})	Average temperature of the thermolysis / K	Crystallite size/ nm		Specific surface area / ($m^2 g^{-1}$)
			d_{104}	d_{110}	
7.3×10^{-6}	3.3×10^{-3}	470	8	34	102
	6.5×10^{-4}	440	–	–	89
	2.2×10^{-4}	–	–	–	76
	1.8×10^{-4}	430	–	–	45
0.49	3.0×10^{-3}	–	10	34	83
	7.0×10^{-4}	490	–	–	58
1.09	7.6×10^{-4}	–	12	34	44

of 7.3×10^{-6} kPa, the thermolysis proceeds via a single step for all the reaction rates used, while at a higher water vapour pressure of 0.49 kPa, the reaction occurs via two steps. The decomposition of FeOOH through an intermediate step, prior to complete dehydration, depends on the partial pressure exerted around the sample by the water vapour generated in the reaction. At the same time, the H_2O pressure not only influences the reaction pathway but also the porosity of the final catalysts.

Figure 7.7 reveals the results of the CRTA preparation of iron catalysts, showing the influence of the reaction environment on the BET surface area of the products. The surface area of the iron oxides increases both with decreasing the controlled level of water vapour pressure and also increasing the reaction rate. Hence, once again the reaction rate and partial pressure of water vapour influence the development the porosity of the catalysts in opposite directions. However, using CRTA, these parameters are under control and it is possible therefore to choose experimental conditions to give a specific pore size distribution and surface area.

When the decomposition of the precursors proceeds at the same controlled constant rate, increasing the overall pressure in the system increases the influence of the water vapour on the system. Chopra et al. [19] showed, that during the

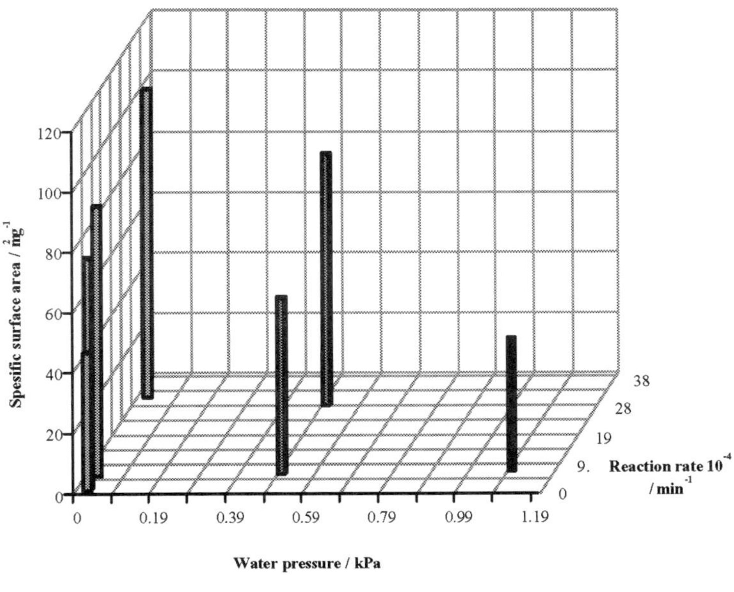

Figure 7.7. Dependence of the specific surface area of the iron catalysts on the CRTA preparation conditions [18].

CRTA decomposition of lipidocrocite (γ-FeOOH), H_2O vapour has a significant catalytic effect on the Fe^{2+} mobility and this promoted surface migration, resulting in the development of the surface. At the same time, the water vapour simultaneously facilitates the growth of pore sizes, pore elimination, an increase in particle size and also the magnetite to hematite transition. Hence, the result of raising the controlled pressure for a certain precursor decomposition rate is an increase in the catalytic effect of water vapour on the crystallisation, which in this case leads to products with larger pore sizes and hence smaller surface areas.

To estimate the effect of temperature on iron oxide preparation, the catalytic effect of H_2O vapour is maintained controlling the residual pressure during the CRTA thermolysis at a constant level for a series of experiments at different constant rates of transformation. In this case, decreasing the decomposition rate results in lowering the temperature at which the process occurs. This is indeed expected, provided the Arrhenius law applies:

$$k = A \exp(-E_a/RT)$$

since it indicates that the lower the rate constant k, the lower the temperature T (A being the "pre-exponential term" and E_a being the energy of activation).

The CRTA control of the reaction environment allows significant variations in the temperature of the overall processes which in turn influences the state of the surface ions. Raising the temperature at which a solid state reaction proceeds increases the mobility of the surface ions involved in the thermolysis and results in the BET surface area of the iron catalysts being characterised by small pore sizes. Moreover, increasing the reaction rates leads to decreasing the time needed to complete the decomposition. If all the other conditions are the same, a faster reaction rate results in a shorter experiment and a decrease of the time allowed for catalytic action of the water vapour, so limiting the growth in pore and leading to higher surface areas.

TEM micrographs of hematite particles prepared using various thermal techniques suggest that, whatever thermal method is used for iron catalyst preparation, the products maintain the size and shape of the precursor, as is expected for topotactic reactions. During the decomposition many pores are generated, resulting in the growth of the surface area. Iron catalysts with a particular surface area are characterised by a specific porosity. Thus, the oxides with high specific surface areas are qualified by slit-shaped pores along the longest direction of the particles, while the catalysts with the lower surface area show randomly distributed isolated circular pores. Considering the slit-shaped micropores are perpendicular to the *c* axis, the (110) peak of the XRD pattern gives an indication of the dimensions of the crystallites in the direction of the pores, while the (104) peak provide information on the direction perpendicular to the pore. The *d* values, calculated from the XRD peaks of the hematite patterns (Table 7.3) show that the dimensions of the crystallites in the (110) direction are identical for all the samples, while the size in the (104) direction decreases with increasing the surface area of the samples [18].

It is not unreasonable to assume, therefore, that the structure and texture of the catalysts can be controlled if the reaction environment is suitably chosen and maintained. For these materials whenever material with a large circular pore is needed to achieve a specific selectivity for a reaction, the catalyst can be produced under a high, controlled, pressure using a slow rate of thermolysis. However, when materials with high surface area and small slit-shaped pores are desirable for a catalytic process, they can be produced using CRTA decomposition with high reaction rates at lower water vapour pressures.

7.1.4. EFFECT OF SELF-GENERATED ATMOSPHERES ON CRTA CATALYST PREPARATION

One of the aims in catalyst manufacture is to minimise the temperature of preparation to reduce both sintering and energy costs. This can be achieved by control of the reaction rate at a reasonably low level (taking into account that a significant decrease in the reaction rate leads to a considerable longer process). A number of SCTA methods result in the temperature being either essentially isothermal or slowly rising through thermolysis. In this way, the temperature of a controlled rate process can be significantly less than that for conventional linear heating. At the same time, gas-phase composition/pressure can often influence the rate of the processes involved. Hence, the temperature of the process cannot be considered in isolation without taking into account the effects of the gas-phase components.

A self-generated atmosphere plays a significant role in thermolyses and ignoring this effect frequently results in an erroneous interpretation of the data. For example, if the gaseous products interact with the solids at the reaction interface, neglecting the reaction atmosphere can lead to contradictory results, as in the case of the decomposition of malachite to form copper oxide catalysts [2]. Some authors report that the temperature of the process rises with increasing partial pressure of CO_2 [20] while the others apparently demonstrate the opposite effect [21]. Other workers provide evidence to suggest that the two product gases are evolved at approximately the same rate throughout the reaction [22], while still others believe that CO_2 and H_2O are evolved at different rates [23]. Yet all of these apparently contradictory conclusions relate to the overall same chemical process, namely:

$$Cu_2CO_3(OH)_2 \rightarrow 2CuO + CO_2 + H_2O$$

To explain such variations, the interpretation of the experimental results requires consideration of the chemical equilibrium and interaction of the reactant and solid products with the product gases, i.e. the catalytic action of evolved water vapour on the crystallisation of CuO and the interaction of the evolved CO_2 with poorly crystalline CuO. Compared with conventional isothermal and linear heating, the CRTA approach provides control of these factors by regulating the self-generated atmosphere and so provides a useful insight into the details of copper catalyst preparation.

The work of Stacey and Shannon [24] on the transformation of hydroxycarbonates such as malachite, aurichalcite and hydrozincite using a SCTA heating routine shows that, in all cases, the initial step of the process nucleates the reaction interface at crystallite surfaces and, as the reactions progress, a porous catalyst structure is formed. In the case of malachite decomposition, the morphology of the product is determined by that of the precursor, while its surface area develops significantly during the thermolysis and reaches a maximum at 50% conversion.

To evaluate the influence of self-generated gases on the formation of catalyst surface area, Reading and Dollimore [23] analysed $Cu_2CO_3(OH)_2$ decomposition using CRTA. The results provide evidence that the process in flowing nitrogen, as well as at reduced pressure, is characterised by a variation in the relative rates of evolution of both CO_2 and water vapour. In the initial stage more H_2O is generated, while at the end the evolution of CO_2 is in excess, so that the two gases are not produced in the constant ratio of the stoechiometric equation. However, the fact that the kinetics of the loss of the two product gases do not follow the same rate and apparently proceed by different mechanisms does not prove that they are evolved independently of one another. It can be suggested that the formation of the surface area of the catalyst in a controlled environment takes place in four stages (Fig. 7.8). The porosity increases dramatically over the early stage of the reaction ($\alpha < 0.2$), after which the major decomposition ($0.2 < \alpha < 0.7$) is characterised by decreasing rate of surface growth. During the next stage decomposition of the precursor ($0.7 < \alpha < 0.9$) the surface development is constant, while at the end ($\alpha > 0.9$) the porous structure of the product is partly destroyed. These features are related to the four stages of the CRTA decomposition seen from the gas evolution and temperature profiles, the first of which is characterised by a higher rate of H_2O evolution compared to that of CO_2 and by a fast temperature increase. The subsequent features are almost isothermal and are distinguished by the crossover point that indicates the reaction interface is richer in carbonate or hydroxide. The last feature is also remarkable by virtue of the relatively high

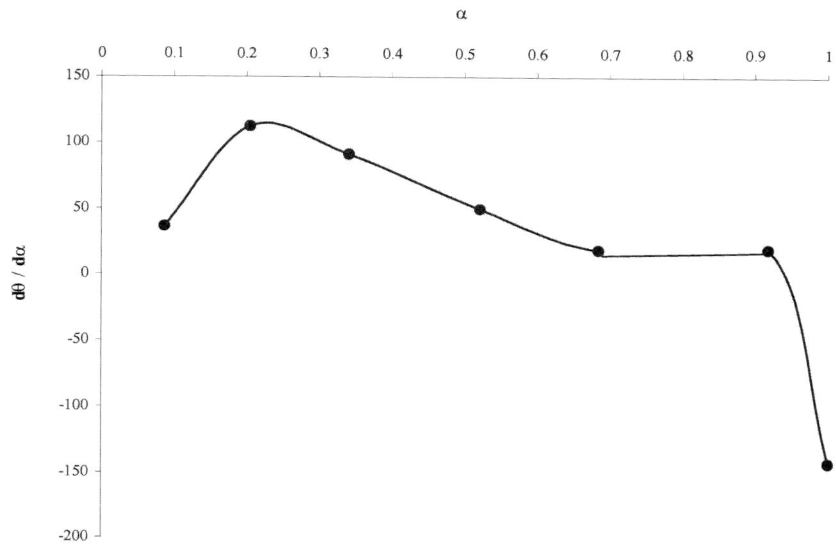

Figure 7.8. Development of the surface area (θ) of a copper catalyst with the extent of the reaction (α) [23].

heating rate where the rate of CO_2 evolution exceeds that for H_2O. These results provide evidence for the loss of water being associated with a greater surface area increase than the loss carbon dioxide. Moreover, the final gas evolution which is also accompanied by the high heating rate even destroys the previously formed structure.

The XRD patterns published by Koga et al. [21] show crystal changes during the CRTA transformation of malachite and demonstrate a similar set of features to the above. The absence of a peak corresponding to CuO for $\alpha < 0.3$ also suggests some difficulty in crystallising the product in the early stages of the reaction. Therefore, the steps of the decomposition of malachite and crystallisation of the product appear to be separated and take place consecutively. During the early stage, when CuO is still poorly crystallised, an interaction between the solid and self-generated carbon dioxide is thought to result in different evolution rates of CO_2 and H_2O. After about half of the malachite transformation, the X-ray diffraction shows the patterns characterising the both reactant and product and further reaction results in pure copper oxide.

The results of Koga et al. [25] on CRTA malachite decomposition in flowing gases (N_2, N_2/H_2O and N_2/CO_2), under vacuum of 4.0×10^{-3} Pa and using quasi-isobaric ambient pressure clarify the influence of self-generated CO_2 and H_2O

on the process. Changing the applied atmosphere while the other conditions, such a mass of the precursor and reaction rate, are kept at the same in all experiments allows clarification of the effect of the self-generated gases. The results show that, whereas all of the decomposition stages proceed at the same constant rate of transformation, the dependence of the extent of the reaction on temperature varies significantly with different atmospheres, which is reflected not just in the temperature ranges but also in the shape of profiles obtained.

Changing the atmosphere of the copper catalyst preparation from vacuum to flowing N_2 and quasi-isobaric conditions of ambient pressure increases the effect of the gases generated during the thermolysis. At the beginning, in absence of the product gases, the temperature required to initiate the decomposition of malachite is lowest in vacuum and rises for flowing gases and quasi-isobaric atmospheres. As the transformation proceeds and the gas evolution becomes significant there are dramatic changes in the temperature regimes of the thermolysis required to control the reaction rate at the same constant level. Under vacuum the temperature continuously increases as the reaction proceeds until ca. $\alpha = 0.3$, after which the decomposition is completed at a lower heating rate in the temperatures range of 890–940 K. The analogous alpha profiles in flowing N_2 and quasi-isobaric atmospheres considerably differ from those in vacuum, having characteristic concave shapes as well as lower temperatures of transformation. A concave CRTA temperature curve [26] indicates a nucleation and growth kinetic mechanism in presence of self-generated atmospheres that is not found in vacuum. After the point of the minimum temperature, the decomposition in both the flowing gas and quasi-isobaric atmospheres is completed in the temperature ranges of 790–840 K and 830–880 K, respectively. Hence, while the thermolysis is initiated at the lowest temperature in vacuum, as the process develops the reaction temperature starts to exceed the others and becomes higher by ca. 100 K and 60 K than in the quasi-isobaric atmosphere and flowing N_2, respectively. It should be noted that the influence of self-generated gases on the decomposition is increased by enlarging of the sample mass. When the rate of transformation is controlled at the same constant level, the temperature of the transformation shifts towards the higher temperatures with decreasing the mass which is consistent with the changes observed at different pressures. Therefore, the excess of the self-generated gases favours lower temperatures for the copper catalyst preparation using the CRTA approach. This tendency agrees with results for conventional linear heating studies on the decomposition of malachite, where the reaction temperature increases as the influence of self-generated atmospheres is decreased by increasing the flow rate of N_2 or reducing the pressure [21].

As mentioned above, the gaseous products obey the different kinetics, so further details on the effect of self-generated gases on the CRTA catalyst preparation can be clarified when the purge gas (N_2) contains an excess of the products CO_2 or H_2O. In all cases the alpha profiles are characterised by concave shapes, but this feature is minor when flowing nitrogen is mixed with CO_2. The overall temperatures of the process are higher by ca. 30 K in flowing N_2/CO_2 and lower by ca. 20 K in N_2/H_2O than found in a nitrogen atmosphere. This shows the unusual influence of CO_2 and H_2O in the reaction environment which, in this work, suggests opposite effects on the process. While the rate of evolution of both gases generated by the reaction itself is maintained constant, the change in the temperature of the decomposition with different atmospheric gases indicates, that from the viewpoint of chemical equilibrium, the normal and opposite effects on the overall kinetics are observed for self-generated CO_2 and H_2O, respectively.

Table 7.4 compares the apparent activation energies for the decomposition of malachite calculated from CRTA curves using the Friedman and Rate Jump CRTA methods. In general, the apparent E_a values increase as the influence of the pressure of the self-generated gases increases, from vacuum to flowing N_2 and quasi-isobaric atmospheres. The result of the composition of flowing gases shows that the activation energy (E_a) increases and decreases very significantly in the presence of excess CO_2 and H_2O, respectively. The lowest value of activation energy for the decomposition in flowing N_2/H_2O provides evidence for the catalytic action of water on the crystallisation of the catalyst, as will be detailed in the following section. On the other hand, the interaction of poorly crystalline CuO with CO_2 results in increasing the activation energy for the thermolysis carried out in flowing N_2/CO_2.

The values of the apparent activation energies of the process in the controlled environment of CRTA calculated using the Friedman and Rate Jump CRTA approaches are in good agreement except for these obtained in a flowing atmosphere. This latter discrepancy arises from the fundamental differences

Table 7.4. The apparent activation energies of the thermal decomposition of synthetic malachite in controlled environment under various atmospheric conditions (averaged over $0.2<\alpha<0.8$) [25]

Atmosphere		Vacuum	Quasi-isobaric	Flowing		
				N_2	N_2/CO_2	N_2/H_2O
E_a / (kJ mol^{-1})	Friedman method	190.8	223.8	192.1	335.0	166.8
	CRTA-RJ method	193.7	226.0	222.4	314.9	155.3

between the two methods. The first requires a series of kinetic curves at different reaction rates, achieved by changing the controlling rate of the process and/or the sample mass, leading to changes in the overall rate of evolution of product gases and mass transfer phenomena affecting the self-generated atmospheric conditions. On the other hand the Rate-Jump CRTA is classified as an isoconversional method [27], in which the different overall rates of the evolution of the product gases before and after the 'rate-jump' are ignored in the estimation of the kinetic parameters. Hence, the values of the apparent activation energies of the process are influenced by the self-generated atmosphere, even in a controlled environment and should be viewed with caution. The master plots of $d\alpha/d\theta$ (θ is surface area) versus α for the CRTA decomposition of malachite are remarkably dependent on the applied atmosphere, indicating the influence of self-generated gases on kinetic parameters as well as on the mechanism the process, especially during the first half of the decomposition. Therefore, the mechanism as well as the temperature regimes of the SCTA process used for the copper catalyst preparation and, hence, the properties of the final product copper catalyst, are significantly influenced by the self-generated atmospheres.

7.2. Redox Reactions Using Temperature and Concentration Control

Solids prepared using simple thermolysis are often not particularly active catalytically. However, the precursors can often be transformed into highly active and selective catalysts through chemical and morphological changes in a final activation stage. Many heterogeneous catalysts are activated by thermally induced redox methods involving isothermal or conventional linear heating in a flow of reactant gas(es). Throughout these processes the inevitable temperature and concentration gradients can result in products with less than optimum properties [28].

The aim of SCTA is to avoid the disadvantages of conventional heating procedures by ensuring a specified reaction rate under particular conditions throughout a thermal process and is reached by controlling the sample heating, as explained in Chapter 3. However, if a process involves reactant(s) in the gas phase, its rate is also affected by the concentration the gas(es). Hence it is possible, in suitable cases, to attain the desired reaction rate using control of the concentration of the reactive gas-phase component(s) rather than the heating. This is the essence of Sample Controlled Reaction Rate by Gas Blending [29] which is described in Chapter 3.

7.2.1. SCTA REDUCTION

During reduction, a precursor metal oxide is activated using hydrogen or carbon monoxide to form fine crystallites of the metal. Excessive temperature during reduction causes sintering, an aggregation of the metal into larger particles that reduces the active surface area. It is also necessary to keep the rate of the process low, as steam, even at low concentrations, is very effective at promoting the sintering of oxides. As a characterisation tool, reduction provides information on the stages and extent of the process, and quantifies the phases present in both the calcined and reduced forms of catalysts. However, if the results are considered without an understanding of the influence of side-effects, they can be very misleading, especially when the reduction gives rise to multiple peaks. Such profiles may indicate the presence of two or more distinct chemical phases in the sample but could also be due to a mixture of particle sizes. Moreover, increasing the heating rate raises the reaction rate, which in turn decreases the concentration of the reducing gas, which can also lead to severe distortions of the resulting reduction profile [3].

Results of Minimising Temperature and Concentration Gradients during the Reduction of Catalysts

As repeatedly pointed out by various authors, conventional TPR results are markedly heating-rate dependent. Furthermore, the results of the linear heating process are complicated by the influence of H_2 consumption rates which vary with both sample mass and reactant gas flow rate. Hence, the resulting profiles

employed which affect both the shape and resolution of the reduction steps as well as peak temperatures. A striking example of such behaviour is shown in the conventional TPR of V_2O_5 at different heating rates (Fig. 7.9). The resulting profiles provide clear evidence that the size, temperature and even the number of reduction peaks observed change very significantly for different heating rates. This arises from variations in the reduction rate as the temperature and concentration of the reducing gas (hydrogen) consequently changes. Furthermore, it is difficult to obtain quantitative information in such cases as the alpha profiles (Fig. 7.10) show inadequate separation of the steps of the process. Moreover, increasing the heating rate distorts the profiles, which, at first sight, may be attributed erroneously to enhanced resolution [30]. However this effect is due largely to the very significant influence of variations in the rate of hydrogen consumption [31].

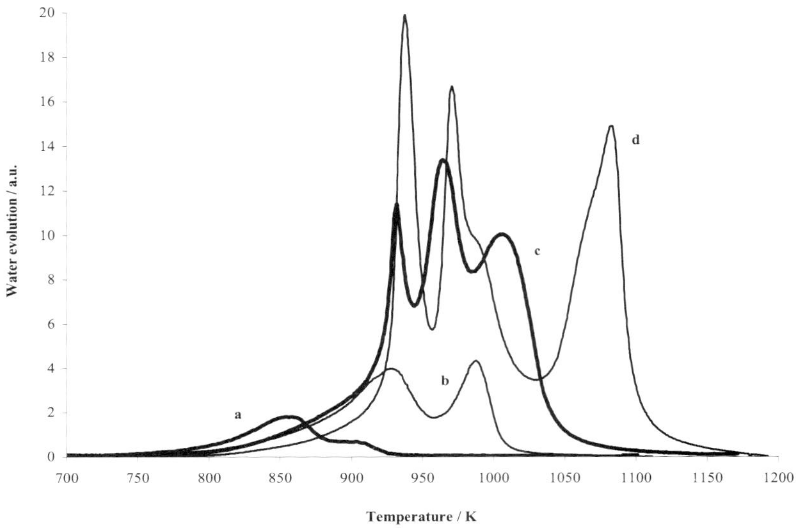

Figure 7.9. Reduction of V_2O_5 using conventional linear heating (a) 1 $Kmin^{-1}$, (b) 3 $Kmin^{-1}$, (c) 10 $Kmin^{-1}$, (d) 15 $Kmin^{-1}$.

Under CRTA conditions, the side-effects arising from temperature and concentration gradients are minimised. In addition, the rate of the reaction can be reduced to a level where the rate-limiting factors are not the availability of hydrogen but the reduction process itself. Hence, using CRTA it is possible to more clearly distinguish overlapping reduction steps to gain more reliable information on the mechanism of the process and obtain better quantitative information [32]. The reduction of various masses of V_2O_5 in controlled environment is shown in Figure 7.11. As the sample mass increases from 10 to 15 and 20 mg, the specific reduction rate (i.e. per unit mass) is decreased, while the same rate of the reduction for all experiments ensures a constant rate of consumption of H_2 despite the change in the sample mass. The alpha curves (Fig. 7.12) show that the reduction of vanadium oxide in this controlled environment gives similar profiles in the temperature domain although the duration of the process varies from one experiment to the other. All the experiments produce similar results with three separate events being clearly distinguished in each. Hence, from the α-curves it is possible to provide substantial evidence for the following reaction pathway:

$$V_2O_5 \rightarrow V_4O_9 \rightarrow VO_2 \rightarrow V_2O_3$$

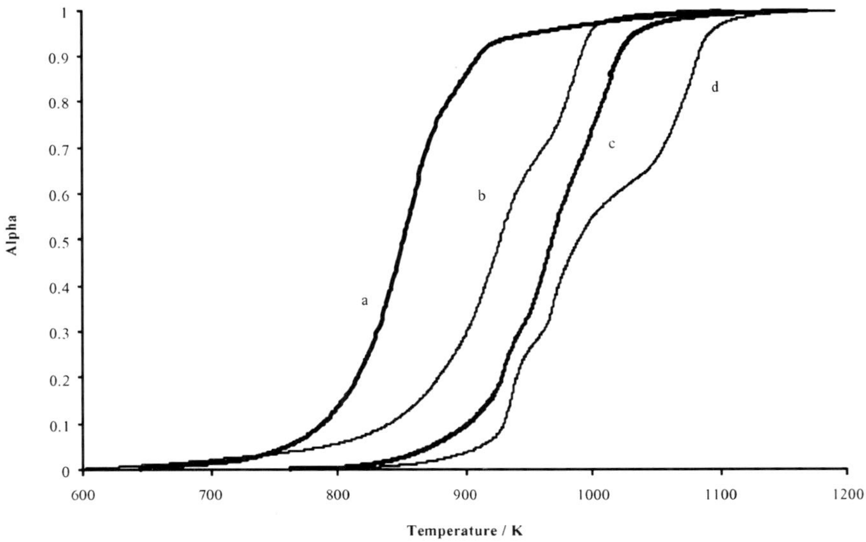

Figure 7.10. Extent of reduction of V_2O_5 using conventional linear heating (a) 1 $Kmin^{-1}$, (b) 3 $Kmin^{-1}$, (c) 10 $Kmin^{-1}$, (d) 15 $Kmin^{-1}$.

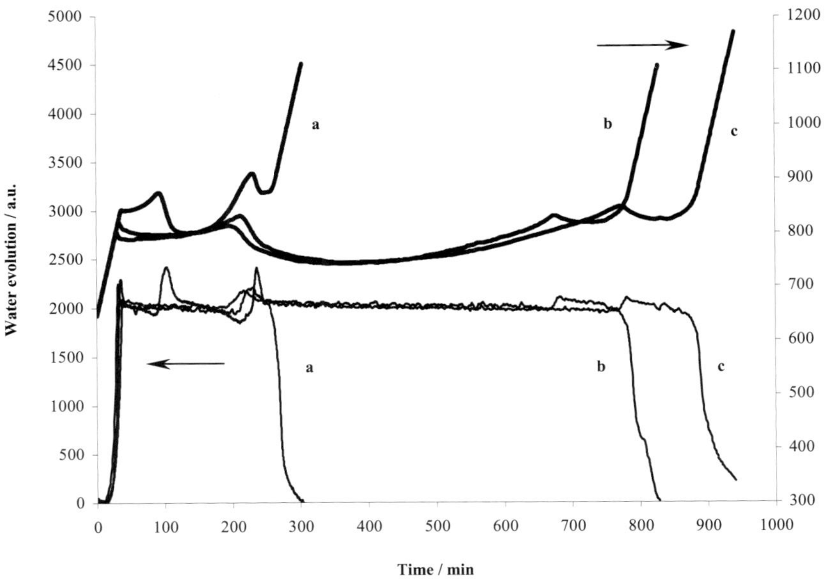

Figure 7.11. Reduction of V_2O_5 using CRTA technique: (a) 10 mg, (b) 15 mg, (c) 20 mg.

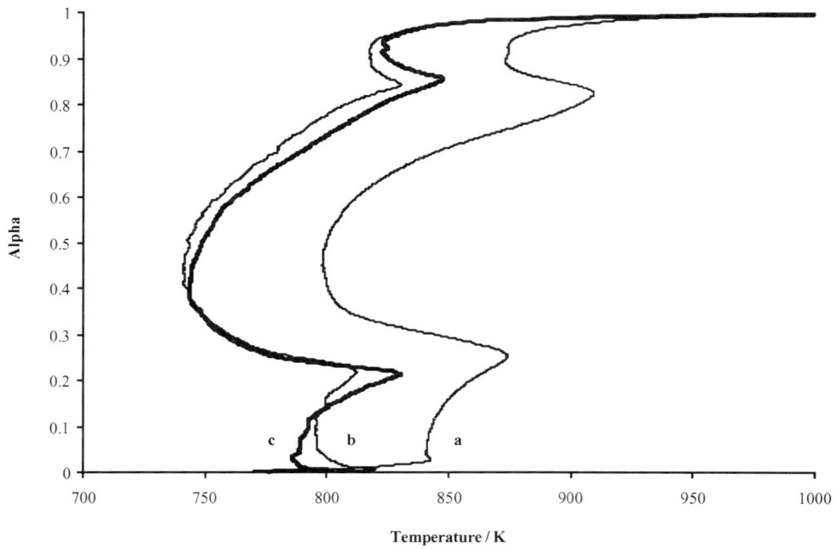

Figure 7.12. Reduction of V_2O_5 using CRTA technique: (a) 10 mg, (b) 15 mg, (c) 20 mg.

The control of the environment during reduction enhances the separation of the consecutive events and, moreover, reveals considerably lower temperatures of reduction than suggested by conventional methods. This is important when the catalyst is required for subsequent testing and the effects of sintering during reduction need to be minimised.

Each of the alpha plots for V_2O_5 reduction exhibits a strong characteristic concave shape. As noted by Criado et al. [33], the shape of the α-temperature curve provides a means of distinguishing between different reaction mechanisms. The CRTA results for vanadium oxide transformation indicate the Avrami-Erofeev kinetic model, i.e. catalyst reduction follows an autocatalytic or nucleation and growth mechanism. The initial 'overheating' during the CRTA experiments provides convincing evidence for a significant nucleation stage. In terms of the reduction of metal oxides, this step can involve the initial removal of oxygen atoms from the lattice. Once the concentration of vacancies reaches a critical value they are then annihilated by lattice rearrangement with the formation of metal nuclei. The nuclei then grow and, as they expand, the area of the sample-product interface, i.e. the interface between the metal oxide and the metal nuclei, begins to increase, causing the reduction process to accelerate. In CRTA experiments, as the reaction attempts to accelerate, the temperature drops in order to maintain it at a constant level as the reaction interface increases. However, when the expanding metal nuclei begin eventually to merge, the area of the sample-product interface decreases and the temperature rises again to prevent deceleration of the reaction. It is this rise, fall and then rise again of the temperature that causes the α versus temperature profile to curve back on itself during the process. In some cases, the metal nuclei formed give rise to autocatalysis in which hydrogen molecules are dissociated to produce the more reactive hydrogen atoms. Such autocatalysis produces similar concave α-temperature profiles [34].

Temperature and Concentration Induced Reduction

Reduction of oxides using certain temperature regimes is well-known and widely applied. However, the rate of reduction is also affected by the concentration of reactant gas(es) as well as the heating rates. Hence, as noted earlier, it is possible to carry out reduction by altering the concentration of the reactive gas-phase component(s) rather than the temperature. Of course, carrying out reduction by altering the concentration of reactant gas(es) is not so flexible as the standard technique in which the temperature is changed. However, choosing a suitable isothermal regime for the reaction allows the successful reduction of oxides simply by changing the H_2 concentration. This novel thermal technique, where the concentration of the reactant gas is linearly increased while the temperature and overall gas flow are kept constant is the 'concentration' analogue of the conventional linear heating method [29].

Figure 7.13 (a, b, c and d, e, f) compares the reduction of copper oxide using linear heating in a blend of hydrogen in helium with a fixed concentration, and a linear gas blending experiment, where the blend of hydrogen in helium is linearly increased at a certain flow rate under isothermal conditions. The resulting water evolution profiles (a, d) are generally similar for both methods, with the H_2O peak being somewhat broader in the linear gas blending run than that for the temperature programmed equivalent, for the reasons described in the introduction to this section.

Furthermore, the reduction can also be successfully carried out in a controlled environment by altering the concentration of the reactant gas(es) instead the temperature using the CRTA approach. Copper oxide reduction, using 'temperature adjustment' SCTA, in a controlled environment proceeds at a significantly lower pre-set reduction rate (ca. 1.4%) compared with a conventional linear temperature programming experiment (Fig. 7.13 (g, h, i)). The results show the expected initial overshoot of the evolved water, after which the reaction rate attains its pre-set constant target level, achieved by changing the temperature. Such behaviour, under constant rate conditions, is typical for the reduction of the oxides and again provides evidence for nucleation and growth process.

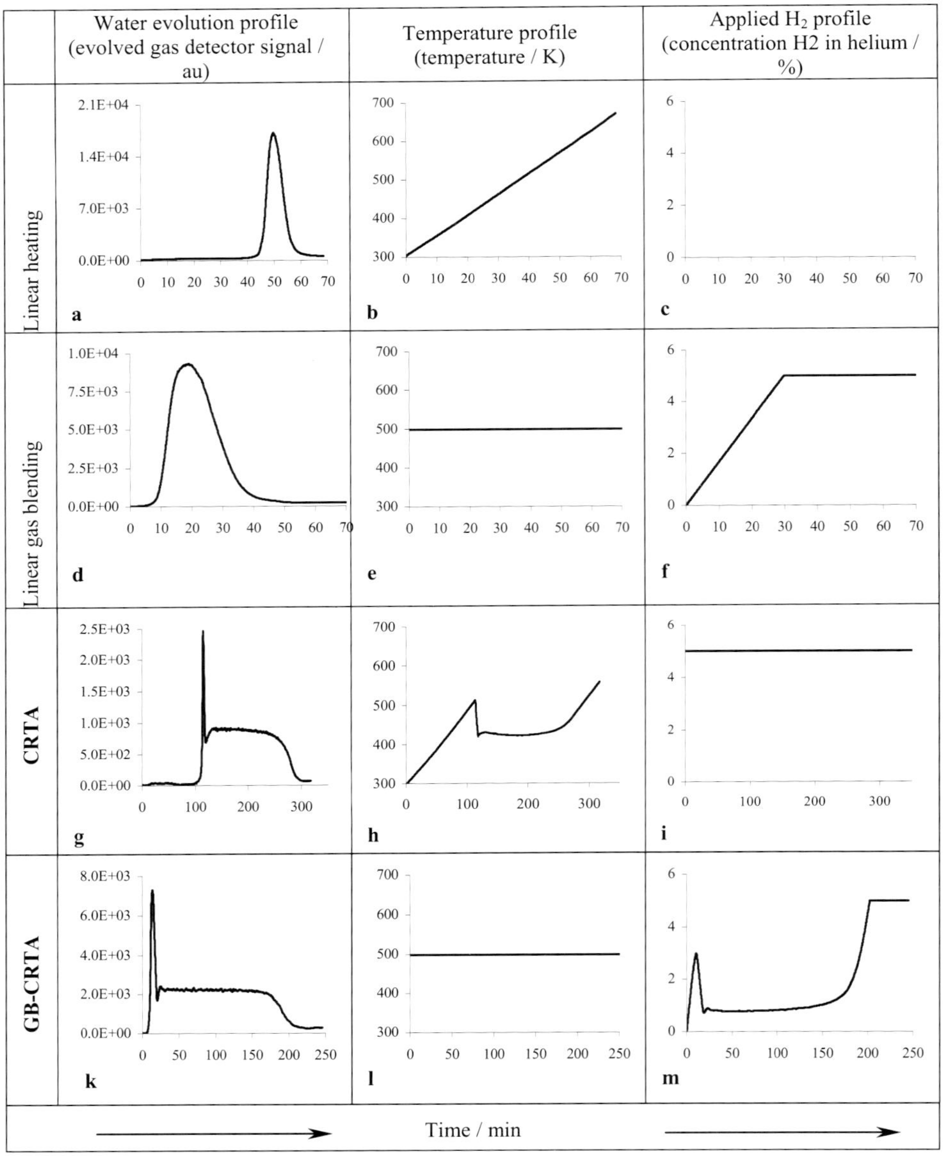

Figure 7.13. Reduction of CuO using (a, b, c) conventional linear heating, (d, e, f) linear gas blending, (g, h, i) CRTA and (k, l, m) GB-CRTA techniques.

In the analogous gas-blending reduction of copper oxide (Fig. 7.13 (k, l, m)), the temperature is held isothermally, while the concentration of the hydrogen gas is altered, as necessary, to maintain the same rate of reaction (water evolution) as in the above CRTA controlled reduction of CuO. The essential similarity of the evolved water profiles in both the temperature and concentration programming experiments (g, k) is apparent, as is the correspondence of the hydrogen concentration profile to that of the temperature (m, h). Analysis of the temperature and gas concentrations involved in the two CRTA experiments shows that a change in either parameter significantly affects the results. In the experiment controlled by temperature, run under 5% hydrogen in helium, an average temperature of ca. 440 K is required to maintain the chosen constant reaction rate. In contrast, the experiment with controlled hydrogen concentration is performed at 498 K and requires an average hydrogen concentration of 1.5% to maintain the same rate of reaction. Thus, for these experiments a change in temperature of 58 K is equivalent to a change in hydrogen concentration of ca. 70%, which is consistent with the general expression for the solid state reactions which shows that, whereas the temperature dependence of the reaction rate is exponential, that of the concentration not. Whilst not quite as simple as to implement as temperature programmed or SCTA controlled process, the gas blending approach has certain advantages in that it enables the dependence of reduction rate on concentration to be evaluated under specific conditions. Both the temperature and concentration controlled CRTA methods maintain a constant environment during the process and so provide all the benefits of the constant rate approach, as discussed earlier.

Activation Energy of Catalyst Reduction in a Controlled Environment

The activation energy is one of the most important fundamental parameters in catalysis but there is much debate in the literature about its validity in terms of its traditional definition [35]. This arises since solid state/surface reactions are influenced not only by chemical considerations but also by physical parameters (e.g. nucleation and nuclei growth, diffusion, sublimation, adsorption and desorption), so the overall rate of the process is the complex sum of a number of individual processes. Their strong dependence on experimental factors such as pressure and temperature, as well as on the sample characteristics, complicates the comparison of E_a values from various sources. In this chapter, this parameter is referred as an 'apparent activation energy' and is used to interpret the overall rate data throughout the complete reaction by applying the Arrhenius equation to each step of reduction. The conventional determination of kinetic parameters usually requires a series of either isothermal or temperature programmed experiments, in which a range of temperatures or heating rates, respectively, is employed. While the isothermal approach has undoubted advantages and has been very widely applied, it is difficult to ensure that the reaction interface and the experimental conditions are identical throughout all of experiments [36]. Linear heating methods, e.g. those of Kissinger, Ozawa and Redhead, not only suffer similarly but are also significantly affected by the usual problems associated with linear heating, as outlined earlier.

CRTA, however, modified to incorporate the 'rate-jump' method, allows the calculation of kinetic parameters from a single experiment. Moreover, the apparent activation energies determined in such a way benefit greatly from minimising the side-effects related to the experimental and sample conditions. In Rate Jump CRTA the rate of reaction is made to alternate between two pre-selected constant values and the changes in sample temperature required to achieve this are monitored, as detailed in a previous chapter. Provided the reaction rate is allowed to reach a constant level at the two successive rates, the corresponding temperature measurements can be used to calculate a value of the apparent activation energy for the reaction by applying the Rate-Jump CRTA form of the Arrhenius equation. This method uses each 'rate-jump' to estimate a value of E_a, allowing any variation in the apparent activation energies during a process to be obtained. Hence, the apparent activation energies so obtained are not only largely independent of the conditions applied, but provide a much greater insight into the processes involved.

The Rate-Jump CRTA data used to estimate the influence of experimental conditions on the apparent value of E_a for the reduction of copper oxide, are shown in Figure 7.14. The overall profiles are similar for all the experiments and reflect the characteristic features of the process, as discussed above. However, in Rate-Jump CRTA after the initial reduction overshoot, typical for CuO, the rest of the experiment differs in that the rate of the process swings between the two pre-set targets of reduction rate, as measured by the rate of water evolution. The magnitude of the 'rate-jump' is the same for all the experiments, ensuring similar, but not necessarily identical, changes in the corresponding temperatures. To provide enough time for achieving a reasonable number of 'rate-jumps', the level of H_2O evolution is lowered as the mass of the sample is decreased [37].

The values of the apparent activation energy obtained for each 'rate-jump' using the Rate-Jump CRTA form of the Arrhenius equation as a function of the extent of reduction (α) are shown in Figure 7.15. There is a significant discrepancy between curves *a-c* and curve *d* that suggests the conditions applied in this experiment do not fully satisfy the assumption that α remains constant for each 'rate-jump', thus changing the reaction interface or mechanism and so influencing the E_a values.

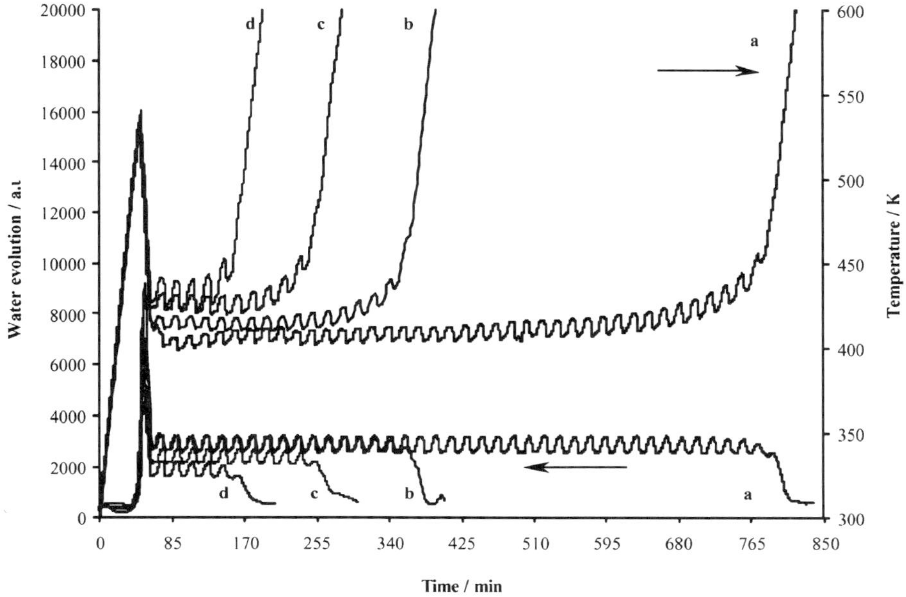

Figure 7.14. Reduction of CuO using the CRTA-RJ technique under the experimental conditions listed below:

Run	Sample mass (mg)	C_1* (mg H_2O min^{-1})	C_2** (mg H_2O min^{-1})	O*** (%)	No. of rate jumps
a	40.0	1.04×10^{-2}	1.26×10^{-2}	0.2	78
b	20.0	1.04×10^{-2}	1.26×10^{-2}	4.3	31
c	10.3	8.10×10^{-3}	1.04×10^{-2}	8.8	21
d	4.7	5.85×10^{-3}	8.10×10^{-3}	15.0	10

*C_1 = lower water evolution rate.
**C_2 = higher water evolution rate.
***O = percentage of the total reduction process which occurred in the initial 'overshoot' of the target reduction rates in each run.

However, if the conditions essential for the Rate-Jump CRTA method are maintained, the activation energies show a similar behaviour, despite the different experimental conditions applied. Overall, the activation energy decreases throughout the course of the reduction, providing evidence for autocatalytic effects. In the range of $0.4 < \alpha < 0.9$ the apparent activation energies determined are essentially identical showing the validity of the Rate-Jump CRTA method. At the same time, there is a significant difference in the

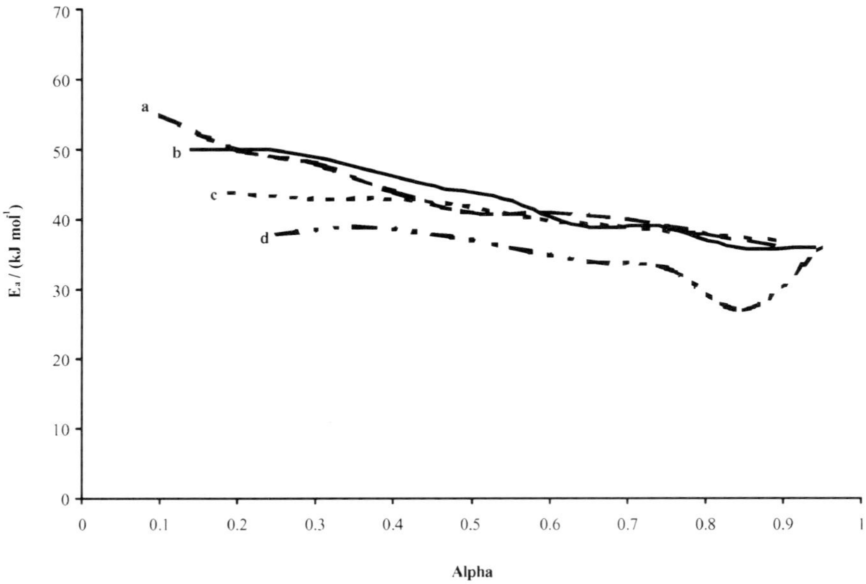

Figure 7.15. Apparent activation energy of CuO reduction using CRTA-RJ technique under conditions listed in the table in Figure 7.20.

initial E_a values. This difference relates to an increase of the percentage of the reaction occurring in the initial or nucleation phase. As the shape of α-temperature plots for CuO reduction indicate a nucleation and growth mechanism, the overall values for apparent activation energies combine both contributions. The activation energy of the nucleation is higher than that required for the growth stage. Thus, if a larger percentage of the reaction occurs in the nucleation part of the CRTA process, the reduction progresses further into the acceleratory growth stage before the 'rate-jumps', and the apparent activation energy measurements, begin. In this case, the initial values of E_a reflect more of the growth stage rather than the nucleation stage and, hence, the apparent activation energies in the beginning of the faster process are lower than that for the slower reaction.

Mechanism of Catalyst Reduction in Conventional and Controlled Environments

Kinetic data for reduction of metal oxides are generally interpreted in terms of nucleation, auto-catalytic or contracting sphere (phase boundary) mechanisms. However, as shown above, a number of experimental factors can produce significant deviations in $f(\alpha)$ and the measured values of A and E_a. In the literature, the reduction of the iron oxides to Fe is claimed to follow various mechanisms and to show a dependence on particle size and the conditions applied. Under conventional conditions, the process follows a phase boundary mechanism (topochemical mode of reaction) or formation and growth of nuclei model (uniform internal reduction) in respect to the large or small particle sizes of the precursor. Furthermore the temperature of the process significantly varies from one source to the other as well, as does the apparent activation energy of the reduction. Depending on the oxide concerned, reduction to Fe can either occur in a single stage (Fe_3O_4) or in two steps (Fe_2O_3). The latter proceeds via magnetite formation where the two processes involved can be consecutive or simultaneous, depending on the conditions:

$$3Fe_2O_3 + H_2 \rightarrow 2Fe_3O_4 + H_2O$$
$$Fe_3O_4 + 4H_2 \rightarrow 3Fe + 4H_2O$$

As the SCTA approach minimises the influence of experimental factors by maintaining the pseudoequilibrium conditions, the kinetic information it provides is more consistent [38]. The reduction in the controlled environment of CRTA (Fig. 7.16) confirms that the transformation of magnetite to hematite is complete prior to the onset of the reaction of Fe_3O_4 to metallic iron and the first step accounts for ca. 11% of the overall process [39]. At the same time, both the steps of SCTA reduction of Fe_2O_3 occur over the same temperature range

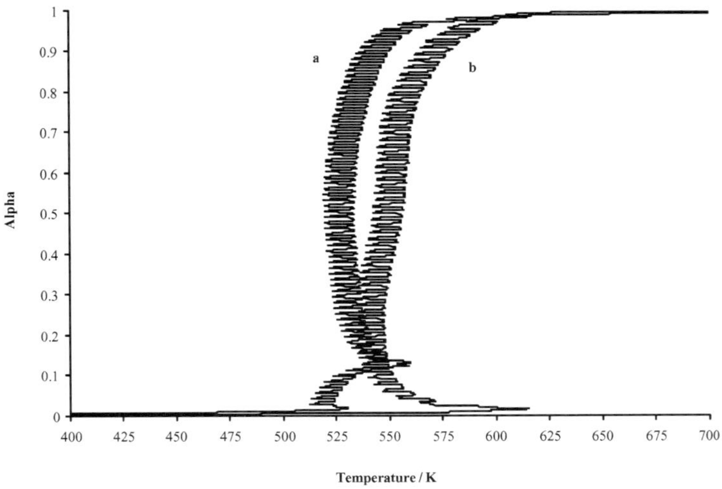

Figure 7.16. Alpha plots for the reduction of Fe_2O_3 (a) and Fe_3O_4 (b) using CRTA-RJ techniques.

(unlike the linear heating process where the difference in T_{max} is ca. 100 K). The apparent activation energy of the reduction Fe_2O_3 to Fe_3O_4 is 96 kJ·mol^{-1}, a value approximately twice that of the transformation of hematite to Fe. The difference in the two parts of the α-temperature curves clearly indicates that the processes are characterised by different kinetics where the first step is attributed to a phase boundary process while the second follows a nucleation/autocatalytic mechanism. It is difficult to show this differences using conventional linear heating reduction, as these to processes overlap under these conditions.

In general, despite the differences in the initial precursor, both the Fe_3O_4 reduction to the final Fe product are characterised by a concave shape of the alpha profiles, which provides strong evidence of a nucleation/autocatalytic mechanism. The reduction temperature increases with decreasing accessibility of the reaction interface since the BET surface areas of Fe_2O_3 and Fe_3O_4 precursors change from 160 to 2 m^{2}g^{-1}, respectively. This is also reflected in the apparent activation energies for the process (Fig. 7.17) as the precursor with the higher surface reduces more easily than the material with low porosity. Although there are significant differences in the surfaces of the precursors and in the reaction pathways, the apparent activation energies in the controlled environment of SCTA differ by less than 10%. Both the processes are characterised by a decrease in the apparent activation energies over the course of the reduction, which is consistent with an autocatalytic effect similar to that the discussed above for the reduction CuO in a controlled environment.

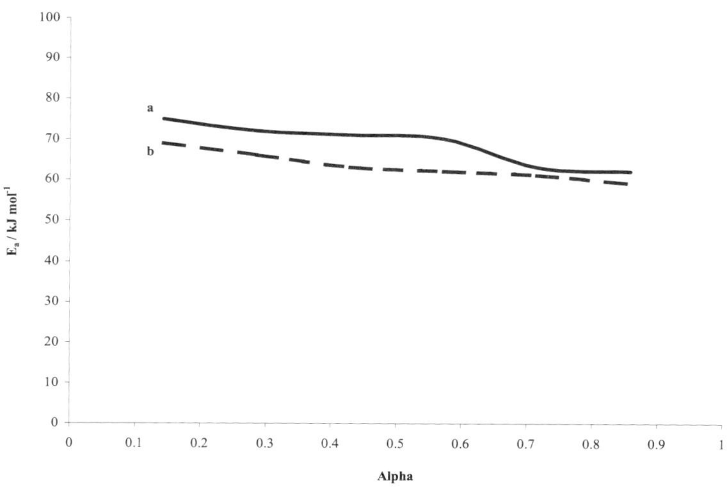

Figure 7.17. Apparent activation energy of reduction of Fe_3O_4 to Fe using Fe_3O_4 (a) and Fe_2O_3 (b) as precursor.

7.2.2. SCTA OXIDATION

Oxidation as well as reduction is used for catalyst preparation and characterisation. In the latter case, the consecutive use of both provides information on the reversibility of catalysts which is crucially important in catalyst preparation and regeneration. In general, oxidation involves passing a stream of oxygen-containing gas through the catalyst bed and its uptake, monitored as a function of temperature, provides valuable information on specific properties of the material. This is an another example of a thermally induced procedure that suffers from temperature and concentration gradients arising during the conventional linear heating. This non-uniformity throughout the linear heating process influences both the quality of the analytical data and the properties of the final products.

Problems of Temperature Control and the Alternative of Gas Blending SCTA. Activation (in this case the development of microporosity), which is a crucial stage in a carbon catalyst and support preparation, is usually carried out by oxidation of the low surface area precursor in steam or carbon dioxide. These processes are endothermic and, therefore, are relatively easy to perform, despite the requirement of high temperatures (1100–1300 K) to drive the reactions:

$$C + CO_2 \rightarrow 2CO$$
$$C + H_2O \rightarrow CO + H_2$$

However, a very significant reduction in activation temperature is possible by carrying out the process in an O_2 atmosphere:

$$C + \frac{1-x}{2} O_2 \rightarrow xCO + (1-x)CO_2$$

In this case, due to the exothermic nature of the reaction, the gasification of carbon is extremely difficult to control and frequently results in violent overheating and excessive burn-off [40]. The conventional linear heating oxidation of carbon at 10 K min^{-1} is characterised by a slow initial reaction rate with a small O_2 uptake and CO_2 evolution (Fig. 7.18). However, the main transformation proceeds within ca. 2 minutes via almost explosive reaction acceleration. Furthermore, within this period the concentration of the reactant O_2 rapidly decreases by ca. 80% (with a corresponding rise in CO_2 evolution). Hence, the complete carbon burn-off takes place under the influence of very large uncontrolled temperature and concentration gradients throughout the sample, giving rise to highly heterogeneous materials and in a large scale process, uncontrolled thermal runaway.

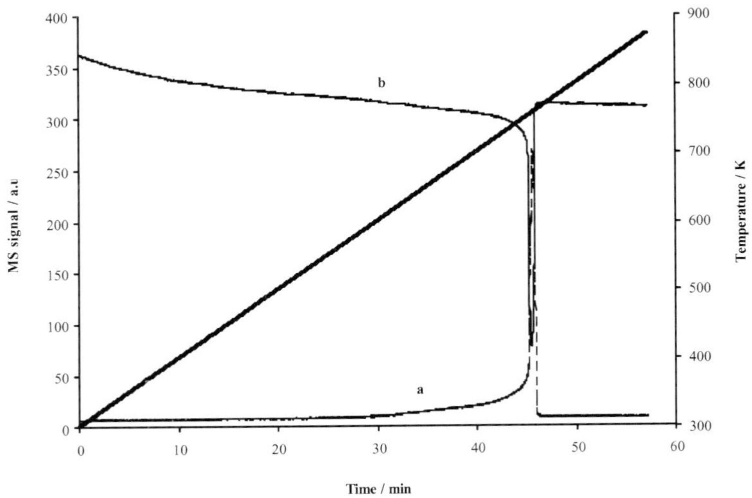

Figure 7.18. Oxidation of an activated carbon using conventional linear heating at 10 $Kmin^{-1}$ ((a) CO_2 and (b) O_2).

Decreasing the reaction rate of the carbon activation can avoid the sudden violent reaction which is exacerbated by the exothermic runaway. Real et al. [41] showed that this is successfully achieved by the application CRTA to carbon oxidation. In a controlled environment, the process occurs at a constant rate that can be as low as desired. In CRTA, the levels of both the product and reactant gases are maintained at a constant level and complete burn off can be achieved with few unwanted the side effects. Moreover, there is a linear relationship between the amount of CO_2 evolved during the activation and the mass lost by the carbon. Hence, an integration of quantity of carbon dioxide formed during the activation in controlled environment enables a pre-determined amount of burn-off to be specified.

The thermal control of the reaction rate improves various aspects of thermal induced processes, as it is repeatedly pointed in this book. However, as noted above, where highly exothermic processes are involved, this method of a control runs into serious problems [42]. The data in Figure 7.19 shows the concentration profiles of reactant O_2 (c) and product CO_2 (a) gases during the oxidation of carbon and the heating regimes (b) required to keep the rate of the process at a constant level using the thermal CRTA approach. Here, the control of the reaction rate at the required level is achieved by changing the temperature of the process, while the level of the reactant gas is the same for all experiments and is maintained constant during the oxidation. The thermal control of carbon oxidation is successful at the lower reaction rates (I and II), which achieve and

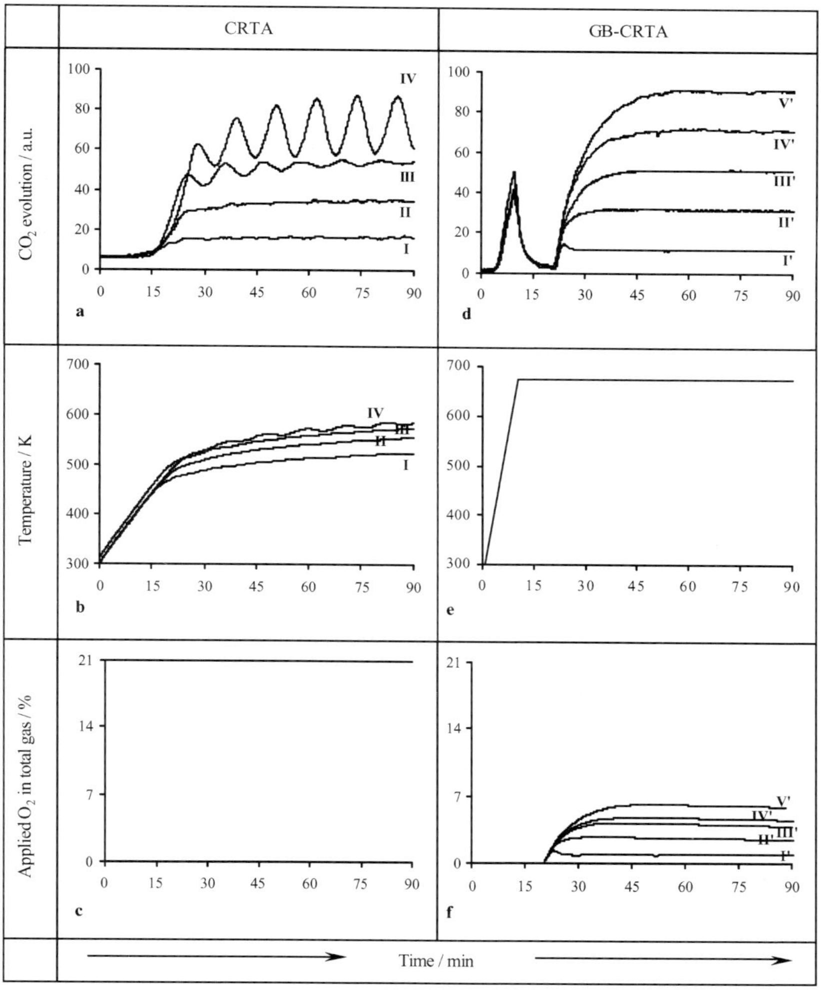

Figure 7.19. Oxidation of carbon using temperature (CRTA) and concentration control (GB-CRTA) ((a, d) reaction rate, (b, e) temperature, (c, f) oxygen input concentration for target reaction rates of (I) 0.15, (II) 0.4, (III) 0.75, (IV) 1.05 and (I') 0.15, (II') 0.35 (III') 0.58, (IV') 0.81, (V') 1.05 mg min^{-1}.

maintain constant rates of CO_2 evolution. At a higher rate (III) an initial oscillation in the rate is observed, which tended to die away with time. However, this oscillation is much greater at the maximum studied rate (IV) and it persisted throughout the experiment, showing the difficulties in maintaining thermal control in such a case.

The corresponding temperature profiles during the CRTA oxidation are also characterised by a marked oscillation during fast carbon burn-off, due to the exothermicity of the carbon-oxygen reaction. At low reactions rates it does not cause a very significant temperature rise, so the furnace temperature does not have to fall in order to maintain a constant rate of CO_2 evolution. However, at higher rates, the temperature rise of the sample is greater and the thermal runaway effect begins so the furnace has to cool slightly to prevent an overshoot of the CO_2 target level. The thermal lag, which arises from the heat capacity of the system, results in the CO_2 level falling below target value causing the furnace temperature to rise again, thus setting up an oscillation in the temperature and rate profiles. Overall, there is a slow net rise in temperature during all the experiments as the most active sites of the carbons appear to react at about 470 K, but the temperature needs to rise continuously to enable the progressively less reactive sites to produce CO_2 at the constant rate specified.

Although it is possible to reduce or eliminate the observed oscillations by careful tuning of the PID parameters in a CRTA experiment with small sample mass, it requires several repeated experiments and is only applicable for a single rate of oxidation. This is because a different rate of oxidation releases more or less energy (in the form of heat) per unit time which, itself affects the rate of the process, thus requiring corresponding alterations in the control parameters. At very high reaction rates the process releases enough heat to cause thermal runaway (i.e. the carbon would ignite) and control via the temperature of an external furnace becomes impossible. Clearly thermal control is opposed by the chemistry of the events. However, "Sample Controlled Reaction Rate by Gasblending" (Sample Controlled Gasblending) employs a system in which the control mechanism is reinforced by the nature of the reaction itself. This removes, to a large extent, many control problems, so enabling the materials to be oxidised at relatively low temperatures. In "Sample Controlled Gasblending" once reaction temperature is reached, control of the process is achieved by controlling the level of oxidant. Any tendency to thermal runaway immediately depletes the level of the oxygen, which reinforced by the control system that simultaneously reduces or chokes off the reactant supply, ensuring rapid and effective control.

The Figure 7.19 (d, e, f) compares the temperature controlled CRTA with alternative technique used the gas blending (Sample Controlled Gasblending) approach. Here, control of the reaction rate at the chosen level is achieved by changing the concentration of supplied oxygen, while the temperature of the oxidation is maintained constant for all experiments in the series. It is notable that in the case of "Sample Controlled Gasblending", the activation of the carbons proceeds under much better control, without any tendency to oscillation. The initial CO_2 evolution peak results from the thermal desorption of carbon dioxide as the sample is rapidly heated to the isothermal level under flowing nitrogen (this is not apparent in the thermal control experiments where it forms part of the CO_2 target level when heating is commenced under flowing air). After this initial period the rate of carbon oxidation accelerates and to rises to the target CO_2 level as the reactant gas is gradually blended into the gas stream. The concentration of O_2 supplied to the sample (Figure 7.19 (f)) shows of a slight fall in the reactant demand as the reaction proceeds at all but the lowest rate. Given that oxygen is always slightly in excess, and that the rate is governed by the number of active carbon sites available, this is due to the development of internal surface area liberating more sites for CO_2 production.

In comparing the two types of control at similar rates of carbon burn-off, it is therefore apparent that it is far easier to achieve a stable reaction rate by controlling the concentration of oxygen at a fixed high temperature, than controlling the temperature in the lower temperature range in a fixed higher concentration of oxygen. In the case of the carbon-oxygen reaction, gas concentration control is effective at any rate of carbon burn-off whereas thermal control is only successful at low reaction rates. Thus, whereas temperature control is unstable for such systems, "Sample Controlled Gasblending" inherently gives rise to improved control of the reaction rate.

7.3. SCTA and Catalysis

The application of heterogeneous catalysts depends on their activity and selectivity that, in turn, is influenced by both the number and nature of reactive surface sites, as well as their accessibility. The structural heterogeneity associated with specific surface area, pore diameter and volume distribution contributes significantly to the adsorption behaviour of these materials. These properties are detailed in the chapter on the application SCTA to adsorption studies, where the authors show the benefits of using the controlled approach rather than conventional TPD. Most, if not all, catalysts are characterised by the heterogeneity of their surface, i.e. the type and amount of active centres, which can be identified by specific features of surface-catalysed reactions. Thermal methods are widely used for this purpose, as the temperature at which a specific event occurs reflects the apparent activation of the rate-limiting step of the surface-catalysed process and so provides information on the reactivity of active centres. However, as the processes at heterogeneous surfaces are complex, control of the environment using the SCTA approach enhances both the quality and accuracy of information obtained [43].

7.3.1. SURFACE-CATALYSED REACTIONS IN A CONTROLLED ENVIRONMENT

Zeolites are widely used porous acid-base materials whose catalytic behaviour depends on the number and nature of the surface functional groups. The interaction of the materials with a specific reactant is often used for characterising the active centres densities of these catalysts. In this case a probe molecule is brought into contact with the material until the system reaches equilibrium. Heating is then applied in order to determine the gas evolution profiles. Alkylamines are a sensitive probe for estimating the acidity of zeolites. For example, isopropylamine desorbs unreacted from Lewis sites but decomposes to propene and ammonia at Brønsted acid sites, via the Hoffman elimination reaction [44]:

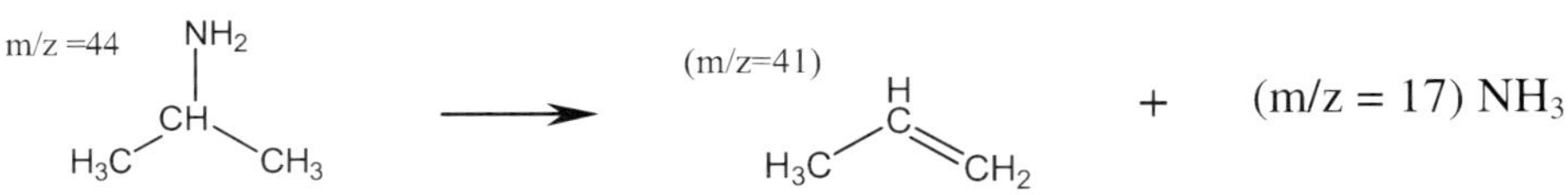

In principle, therefore, an analysis of the thermal induced gas evolution profiles gives an indication of the strength and number of the active acid sites present.

The Na^+ form of Y zeolite possesses only Lewis acidity and conventional linear heating of the equilibrated system results in a simple single peak due to isopropylamine evolution which desorbs unreacted from the surface (Fig. 7.20). Such profiles give well-defined qualitative information. However, if quantitative data are required, the influence of temperature and pressure gradients on diffusion, especially in the case of porous materials such zeolites, must be considered [45]. As is well known, the rate of these processes varies significantly with changing heating rates. Furthermore, the concentration of the desorbed species inside the porous solid, for a given fraction of the process (alpha), differs remarkably for low and high heating rates. Moreover, as the rate of the process changes significantly throughout the course of conventional TPD, mass transfer effects in porous materials lead to continuously varying concentration gradients and/or pressure build-ups in the sample [46]. In contrast, in the controlled environment of Rate Jump CRTA the underlying rate of isopropylamine evolution from the solid-gas interface is low and constant (Fig. 7.21), being maintained by continuously variable heating rate of ca. 0.2 $K min^{-1}$. In this way problems of diffusion limitation throughout the process are minimised. Moreover, using the CRTA approach, both the rate of gases evolved from the catalyst bed and the residual pressure during the process can be adjusted in such a way as to eliminate the effects of any diffusion restrictions [47].

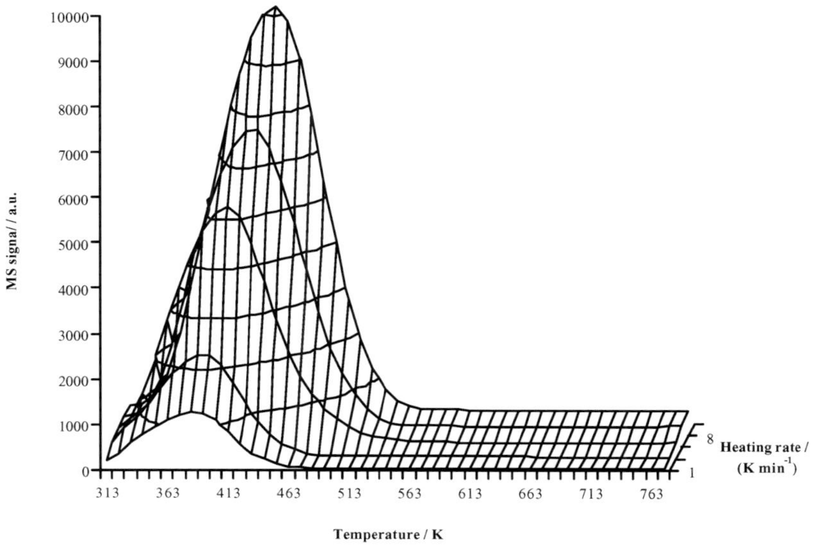

Figure 7.20. Conventional TPD of isopropylamine from NaY zeolite using heating rates at 1, 2, 4, 8 and 16 $Kmin^{-1}$.

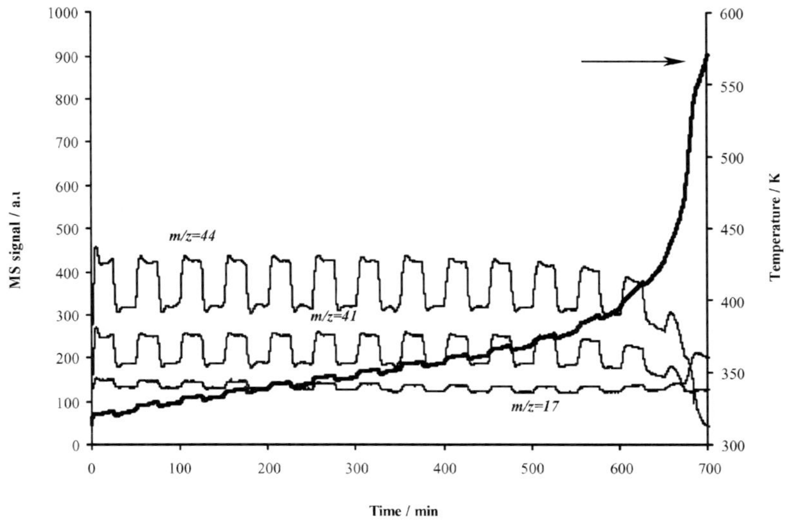

Figure 7.21. Desorption of isopropylamine from NaY zeolite using CRTA-RJ techniques: (*m/z*=44) isopropylamine, (*m/z*=17) ammonia and (*m/z*=41) propene.

In contrast to NaY, the H^+ form of a Y zeolite exhibits both Brønsted and Lewis acidity. As expected therefore, the profile of gas evolution resulting from the isopropylamine-catalyst interaction using conventional linear heating is complicated by the presence of additional peaks of propene ($m/z = 41$) and ammonia ($m/z = 17$) (Fig. 7.22). The low-temperature feature arises from the physical desorption of isopropylamine from weak acid sites, defect sites in the zeolite structure, or hydrogen bonding of amines which are themselves protonated by the Brønsted sites. The high temperature peaks result from the evolution of propene and ammonia due to the surface-catalysed reaction of the amine at the strong acid sites in the HY zeolite, a feature associated with molecules which are protonated at Brønsted sites. In this case, using conventional linear heating causes not only problems in obtaining qualitative data but also makes it very difficult to get quantitative information (unless a mass spectrometer is used), as the two processes, i.e. the desorption of isopropylamine from Lewis sites and the Hoffman elimination reaction, strongly overlap in the temperature region 520–670 K.

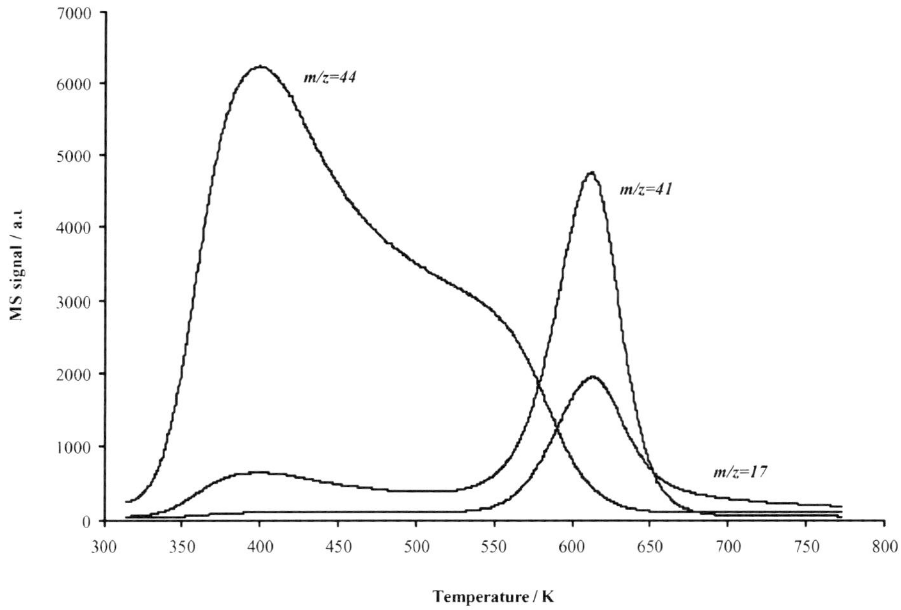

Figure 7.22. Conventional TPD/TPR ar HY zeolite using a linear heating rate of 10 Kmn^{-1}: (m/z=44) isopropylamine, (m/z=17) ammonia and (m/z=41) propene.

Under SCTA conditions, the underlying rate of the first event, i.e. the physical desorption of unreacted adsorbate, can be kept low enough to allow it to reach virtual completion before the consequent surface-induced reaction starts, making it possible to separate it from the other surface processes occurred at the HY zeolite (Fig. 7.23). Unlike the physical desorption isopropylamine from the NaY zeolite, where the profiles of the detected *m/z* fragments are essentially similar, the probe evolution from the acidic form of the zeolite is characterised by a range of different *m/z* signals. It can be seen that the fragment *m/z* = 41 is controlled at a constant level throughout both the desorption and reaction, but it originates from two different processes, initially the physically desorbed probe and finally from the product of the surface reaction. Hence, the first section of this trace arises from physical desorption (here *m/z* = 41 is a fragment of isopropylamine) while the second corresponds to the surface reaction (where *m/z* = 41 is now a fragment of propene), as can be seen from the marked decrease in the physically desorbed amine (*m/z* = 44) and the simultaneous production of ammonia (*m/z* = 17) at the change-over point. Each of the two processes is characterised by its own the specific temperature profile that clearly distinguishes the two events. The temperature regime provides

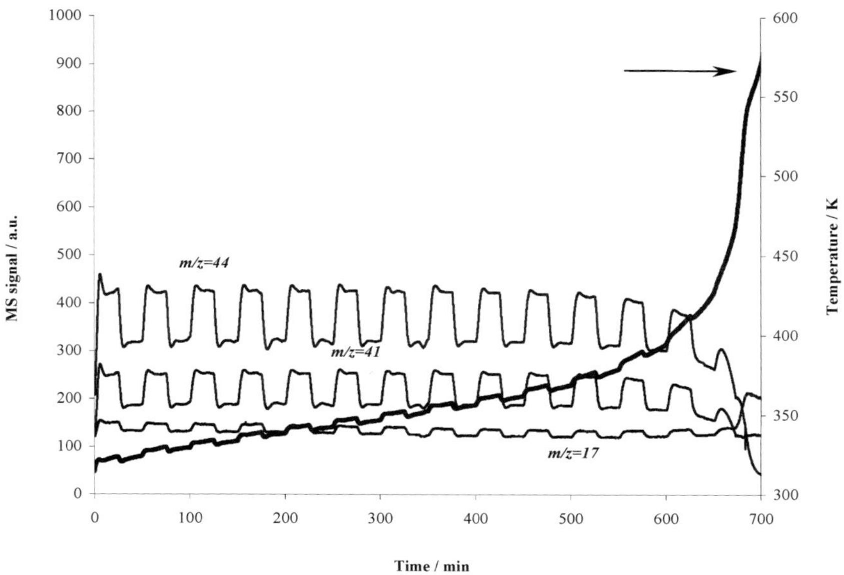

Figure 7.23. Desorption and reaction at HY zeolite using CRTA-RJ technique: (*m/z*=44) isopropylamine, (*m/z*=17) ammonia and (*m/z*=41) propene.

evidence that the average heating rate required to maintain both processes at the same level is about ten times higher for the isopropylamine desorption (ca. 0.7 K min^{-1}) compared with the catalytic reaction (ca. 0.07 K min^{-1}). Hence, the gas evolution profiles from the HY zeolite, especially in region of simultaneous desorption and reaction, using conventional TPD/TPR (linear heating rates) are complicated by diffusion effects which are caused not only by temperature and pressure gradients but also by the different rates of the two processes. However, in CRTA, the products of the desorption and reaction are forced to follow the same rate of evolution, which can be set as low as necessary for a given microporous structure, so that any differences in kinetics can be attributed to the fundamental nature of the processes, by minimising the influence of diffusion [48].

2-methyl-3-butyn-2-ol is a more sensitive probe than isopropylamine as it reacts selectively over a wider range of catalyst functionalities, and so is used to probe surface heterogeneity [49]:

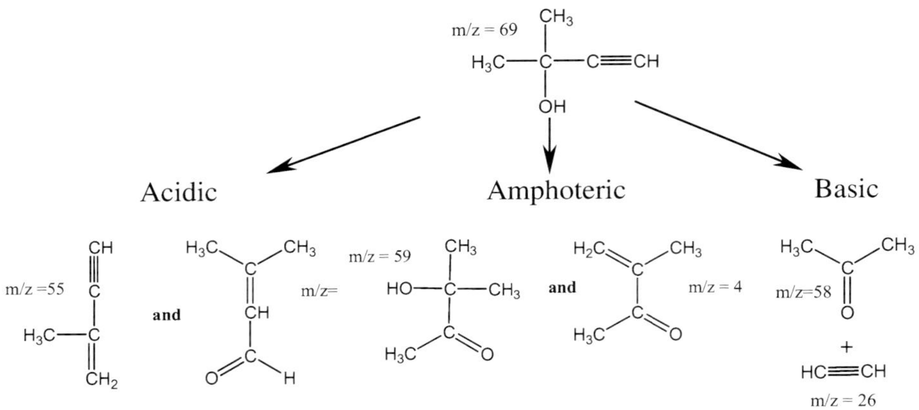

This example is complicated as the reaction environment may contain a mixture of the unreacted probe and at least six products, and each of the fragments can be attributed to more than one substance. However, with ca. 80% accuracy, the *m/z* values shown above characterise a particular product. The analysis of the mass spectra of the mixture suggests that the *m/z* = 39 fragment can be used to control the rate of the overall loss of the 2-methyl-3-butyn-2-ol from the gas-solid interface, as this fragment is present in the fragmentation pattern of all of the products. In a conventional linear heating procedure (Fig. 7.24), two principal reactions occur at the acid and amphoteric sites, which are characterised by different reaction rates, and, hence, diffusion laws. The products of the reactions catalysed by acid (*m/z* = 66 for 2-methyl-1-

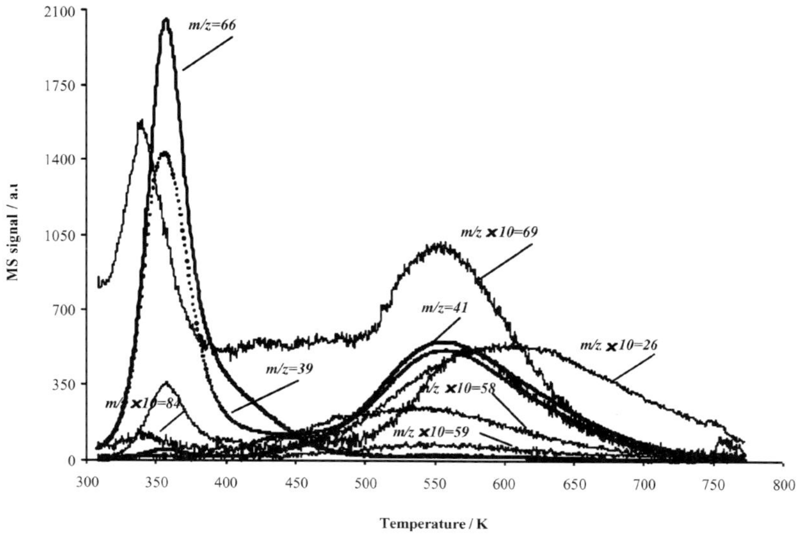

Figure 7.24. TPR of 2-methyl-3-butyn-2-ol at HY zeolite using heating at 10 $Kmin^{-1}$: *m/z*=26 (acetylene), *m/z*=39 (controlled fragment), *m/z*=41 (3-methyl-3-butene-2-one), *m/z*=58 (acetone), *m/z*=59 (3-hydroxy-3-methyl-2-butanone), *m/z*=66 (2-methyl-1-buten-3-yne, *m/z*=69 (2-methyl-3-butyn-2-ol) and *m/z*=84 (prenal).

buten-3-yne and *m/z* = 84 for prenal) and amphoteric (*m/z* = 41 for 3-methyl-3-butene-2-one and *m/z* = 59 for 3 hydrohy-3-methyl-2-butanone) sites are simultaneously evolved in the temperature region of ca. 370–520 K. The features associated with the presence of the active basic sites in this case result in the evolution of acetone (*m/z* = 58) and acetylene (*m/z* = 26). Here they are insignificant but as they arise at the same temperatures as the reaction at the amphoteric sites, they may possibly be attributed to these surface group.

The zeolite catalysed reactions of 2-methyl-3-butyn-2-ol in a controlled environment at the overall constant and low reaction rate (ca. 2% compared with the probe evolution in the corresponding linear heating procedure at 10 K min^{-1}), achieved by maintaining an underlying constant level of the *m/z* = 39 fragment, shows a clear separation in the temperature of the two processes (Fig. 7.25). Moreover, there is the significant (and reproducible) temperature variation at ca. 540 K, accompanied by an increase in the signal of *m/z* = 26 that in this system at 85% is attributed to acetone, the product of the reaction catalysed by basic sites. This feature does not appear in the conventional linear heating procedure, demonstrating the sensitivity of SCTA techniques towards this catalytic reaction. Hence, the reaction of 2-methyl-3-butyn-2-ol on the HY zeolite in a controlled environment is noteworthy not just because of the

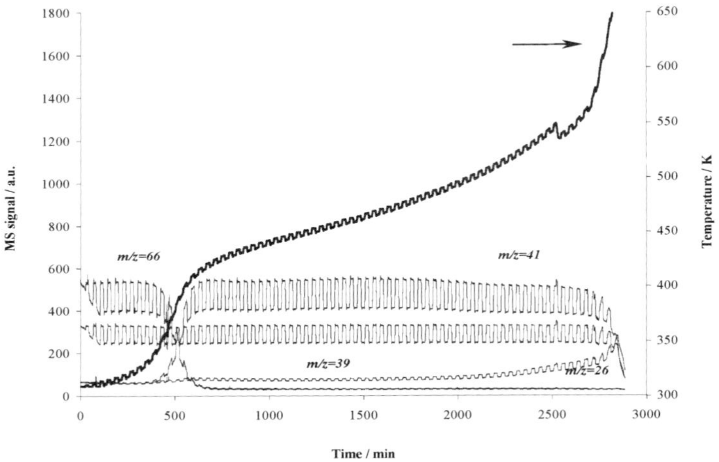

Figure 7.25. Reaction of 2-methyl-3-butyn-2-ol at HY zeolite using CRTA-RJ techniques: (*m/z*=26) acetylene, (*m/z*=39) controlled fragment, (*m/z*=41) 3-methyl-3-butene-2-one and (*m/z*=66) 2-methyl-1-buten-3-yne.

decreased influence of temperature and concentration gradients that result in an enhanced resolution but also because of the additional information obtained on the reactivity of catalytic centres.

The alpha plots for both catalytic reactions (isopropylamine and 2-methyl-3-butyn-2-ol) on the HY zeolite obtained using conventional linear heating are remarkably different from those found in a controlled environment (Fig. 7.26). The SCTA results are characterised by lower temperatures of reaction and pseudo-isothermal section of the processes. There is also a significantly enhanced resolution of events that allows each process to be completed in a narrow specific temperature interval that gives a much better separation of the products of the catalytic reactions.

7.3.2. ENERGETICS OF SURFACE ELIMINATION PROCESSES

The activation energies of surface processes characterise the reactivity of the different functional sites and therefore provide invaluable information on their catalytic activity and selectivity. Some of the methods used in related studies, i.e. microcalorimetery, successfully estimate acidity of solid catalysts by measuring the energy of adsoption of a probe molecule. A significant advantage of thermally induced desorptions and reactions is that they reveal the actual reactivity of the surfaces as they evaluate not simply a strength of functional groups but their real catalytic response.

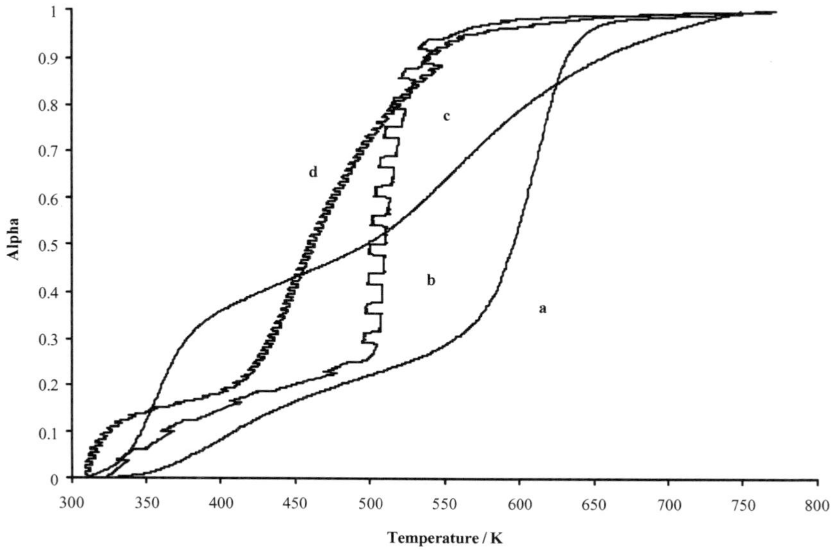

Figure 7.26. Extent of surface induced reactions of isopropylamine (a,b) and 2-ethyl-3-butyn-2-ol (c, d) on HY zeolite using conventional linear heating (a, c) and CRTA-RJ (b, d) techniques.

According to the generally accepted theory of solid state kinetics, the shift in the peak maximum of evolved gases, in linear heating experiments, towards higher temperatures with increasing heating rates (e.g. Fig. 7.26) can be used to calculate the apparent activation energy by applying the Redhead form of the Arhenius equation. Apparent activation energies of the above surface induced processes on the Na^+ and H^+ forms of Y zeolite, as calculated using the results of the linear heating technique are shown in Table 7.5. The value of E_a for the same processes can be also estimated using the SCTA approach. The profiles required for the calculation of the apparent activation energies in the controlled environment are shown in Figures 7.21, 7.23 and 7.25. For each 'rate-jump', a value for E_a is obtained using the CRTA form of the Arhenius equation. A great advantage of the Rate Jump CRTA method is that any variation in the value of the apparent activation enegries with extent of the process are determined. This is shown in Figure 7.27 where the values of E_a are displayed as a function of the surface coverage of the active sites being studied (assuming that the extent of the process (α) is equal to ($1-\Theta$) where Θ is the fraction of surface coverage of these sites). In this way information about distribution of the strengths, or more accurately the reactivities, of catalytic sites can be gained [48].

Table 7.5. Apparent activation energies, E_a (kJ mol^{-1}), for surface elimination processes on the Na^+ and H^+ form of theY zeolites calculated using a conventional linear heating method

<table>
<tr><th colspan="2">Functional groups</th><th colspan="2">HY</th><th>NaY</th></tr>
<tr><td>Base</td><td rowspan="2"></td><td>178</td><td>–</td><td>–</td></tr>
<tr><td>Amphoteric</td><td>112</td><td>–</td><td>–</td></tr>
<tr><td rowspan="2">Acid</td><td>Strong</td><td rowspan="2">75</td><td>104</td><td>–</td></tr>
<tr><td>Weak</td><td>64</td><td>60</td></tr>
</table>

At the same time as the Rate Jump CRTA method provides more detailed information on apparent activation energies, their average values differ from those estimated using conventional linear heating methods (Table 7.5 and Fig. 7.27). It should be pointed out again that the results discussed in this section are obtained in a custom-designed mass spectrometer solid insertion probe (SIP) that is specifically built to study very small sample masses in high vacuum so as to minimise the influence of temperature and concentration gradients during conventional linear heating. However, the use of the SIP alone does not entirely eliminate the problems of the linear heating techniques, as is seen in the E_a values.

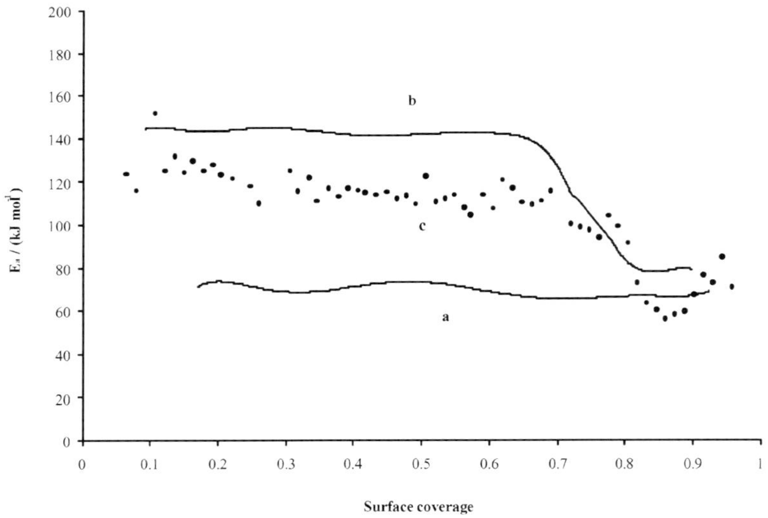

Figure 7.27. Activation energies of surface processes of isopropylamine on NaY (a) and HY (b) zeolite and 2-ethyl-3-butyn-2-ol on HY (c) calculated using CRTA-RJ approach.

During conventional linear heating TPD/TPR_x the rate of the process has a maximum whose magnitude increases with increasing heating rate. Therefore, at higher heating rates the concentrations of the desorbed species inside the porous solid are significantly greater than found with low heating rates, so the diffusion of gases from the micropores may well be rate determining and the desorption/reaction of the probe molecule is no longer the observed rate limiting step. Thus, the internal pressure build-up affects the rate of the product removal, leading to the observed reaction rate giving rise to errors in the value obtained for the apparent activation energy using conventional linear heating. In contrast, the reaction rate in the Rate Jump CRTA experiment can be as low as required, compared with the maximum obtained during the linear heating runs. Hence, the effects of diffusion, outlined above, become insignificant, giving a more accurate estimate of the apparent activation energies. Furthermore, the influence of the enthalpy changes on diffusion, desorption and reaction also cannot be ignored and their effects again vary from one heating rate to another. However, the use of low reaction rates during Rate Jump CRTA experiments provides more time for the system to accommodate these changes and so reduces the local variations in temperature thus caused, again providing better conditions for measuring E_a accurately.

As noted above, unlike conventional linear heating that estimates the overall energetics of surface functional groups, the Rate Jump CRTA approach provides additional information on the distribution of the these sites in the catalyst. Figure 7.27 shows the values of the apparent activation energies as a function of the extent of the processes occurring at the surface of the two zeolites. The curves a and b clearly show the variation in the values of E_a for the catalytic reaction on the HY zeolite that is attributed to the distribution of the different functional sites at the catalyst surface. The 2-methyl-3-butyn-2-ol reactions on the HY zeolite provide evidence that the majority (ca. 60%) of the active groups of the catalyst are amphoteric, with uniform reactivity. The rest of the active centres of the zeolite are basic (ca. 10 %) and acidic (ca. 20%). The latter, in turn, can be estimated to be ca. 30 % weak (Lewis) and ca. 70 % strong (Brønsted), using isopropylamine as the test reaction for the HY zeolite. The strong acid sites (coverage from 0.1–0.65) are uniform in strength while the apparent activation energy of desorption of the probe from weak sites of HY, over the range of coverage from 0.8 to 0.9, is lower and reasonably similar to the value determined for the Na^+ form of the zeolite. The NaY zeolite shows no strong acid sites present and the apparent activation energy of the isopropylamine desorption from NaY, measured from a surface coverage 0.18 to 0.9, is constant through the entire process. Hence, the application SCTA to catalytic reactions can provide a detailed insight into the nature and distribution of active sites in heterogeneous catalysts with a complex surface chemistry.

References

1. Weisz, P.B. and Hicks, J.S. The behaviour of porous catalyst particles in view of internal mass and heat diffusion effects (Reprinted from Chem. Engng. Sci., Vol. 17, p. 265–275, 1962). Chemical Engineering Science 50, 3951–3958 (1995).
2. Tiernan, M.J., Barnes, P.A. and Parkes, G.M.B. Use of solid insertion probe mass spectrometry and constant rate thermal analysis in the study of materials: Determination of apparent activation energies and mechanisms of solid-state decomposition reactions. Journal of Physical Chemistry B 103, 6944–6949 (1999).
3. Thomas, J.M. and Thomas, W.J. Principles and practice of heterogeneous catalysis (VCH, Weinheim, New York, Basel, Cambridge, Tokyo, 1996).
4. Barnes, P.A., Parkes, G.M.B. and Charsley, E.L. High-performance evolved gas-analysis system for catalyst characterisation. Analytical Chemistry 66, 2226–2231 (1994).
5. Rouquerol, J. Controlled rate evolved gas analysis: 35 years of rewarding services. Thermochimica Acta 300, 247–253 (1997).
6. Alcala, M.D. et al. Constant Rate Thermal Analysis (CRTA) as a tool for the synthesis of materials with controlled texture and structure. Journal of Thermal Analysis and Calorimetry 56, 1447–1452 (1999).
7. Sorensen, O.T. Quasi-Isothermal Methods in Thermal-analysis. Thermochimica Acta 50, 163–175 (1981).
8. Sorensen, O.T. RCTA techniques used in studies of solid state reactions in inorganic compounds. Journal of Thermal Analysis and Calorimetry 56, 17–26 (1999).
9. Parkes, G.M.B., Barnes, P.A., Charsley, E.L., Reading, M. and Abrahams, I. Real-time analysis of peak shape: a theoretical approach to sample controlled thermal analysis. Thermochimica Acta 354, 39–43 (2000).
10. Parkes, G.M.B., Barnes, P.A. and Charsley, E.L. New concepts in sample controlled thermal analysis: Resolution in the time and temperature domains. Analytical Chemistry 71, 2482–2487 (1999).
11. Ozawa, T. Temperature control modes in thermal analysis. Journal of Thermal Analysis and Calorimetry 64, 109–126 (2001).
12. Paulik, F., Paulik, J. and Arnold, M. Investigation of the Phase-Diagram For the System Ni(NO3)2-H2O and Examination of the Decomposition of Ni(NO3)2.6H2O. Thermochimica Acta 121, 137–149 (1987).
13. Llewellyn, P.L. et al. Preparation of reactive nickel oxide by the controlled thermolysis of hexahydrated nickel nitrate. Solid State Ionics 101, 1293–1298 (1997).
14. Criado, J.M., Ortega, A. and Real, C. Mechanism of the Thermal-Decomposition of Anhydrous Nickel Nitrate. Reactivity of Solids 4, 93–103 (1987).
15. Gotor, F.J., Perez-Maqueda, L.A., Ortega, A. and Criado, J.M. Kinetic analysis of solid state reactions by means of stepwise isothermal analysis (SIA) and

constant rate thermal analysis (CRTA) – A comparative study. Journal of Thermal Analysis and Calorimetry 53, 389–396 (1998).
16. Naono, H., Nakai, K., Sueyoshi, T. and Yagi, H. Porous Texture in Hematite Derived From Goethite-Mechanism of Thermal-Decomposition of Goethite. Journal of Colloid and Interface Science 120, 439–450 (1987).
17. Perez-Maqueda, L.A., Criado, J.M., Subrt, J. and Real, C. Synthesis of acicular hematite catalysts with tailored porosity. Catalysis Letters 60, 151–156 (1999).
18. Perez-Maqueda, L.A., Criado, J.M., Real, C., Subrt, J. and Bohacek, J. The use of constant rate thermal analysis (CRTA) for controlling the texture of hematite obtained from the thermal decomposition of goethite. Journal of Materials Chemistry 9, 1839–1845 (1999).
19. Chopra, G.S. et al. Factors influencing the texture and stability of maghemite obtained from the thermal decomposition of lepidocrocite. Chemistry of Materials 11, 1128–1137 (1999).
20. Henmi, H., Hirayama, T., Mizutani, N. and Kato, M. Thermal-decomposition of basic copper carbonate, CuCO3.Cu(OH)2.H2O, in carbon-dioxide atmosphere (0–50 Atm). Thermochimica Acta 96, 145–153 (1985).
21. Koga, N., Criado, J.M. and Tanaka, H. Apparent kinetic behavior of the thermal decomposition of synthetic malachite. Thermochimica Acta 341, 387–394 (1999).
22. Brown, I.W.M., Mackenzie, K.J.D. and Gainsford, G.J. Thermal-decomposition of the basic copper carbonates malachite and azurite. Thermochimica Acta 75, 23–32 (1984).
23. Reading, M. and Dollimore, D. The application of constant rate thermal-analysis to the study of the thermal-decomposition of copper hydroxy carbonate. Thermochimica Acta 240, 117–127 (1994).
24. Stacey, M.H. and Shannon, M.D. The decomposition of Cu-Zn hydroxy-carbonate solid solutions. Material Science Monographs 28, 713–718 (1985).
25. Koga, N., Criado, J.M. and Tanaka, H. Kinetic analysis of the thermal decomposition of synthetic malachite by CRTA. Journal of Thermal Analysis and Calorimetry 60, 943–954 (2000).
26. Ortega, A. CRTA or TG? Thermochimica Acta 298, 205–214 (1997).
27. Reading, M. The kinetics of heterogeneous solid state decomposition reactions; a new way forward? Thermochimica Acta 135, 37–57 (1988).
28. Tiernan, M.J., Fesenko, E.A., Barnes, P.A., Parkes, G.M.B. and Ronane, M. The application of CRTA and linear heating thermoanalytical techniques to the study of supported cobalt oxide methane combustion catalysts. Thermochimica Acta 379, 163–175 (2001).
29. Parkes, G.M.B., Barnes, P.A. and Charsley, E.L. Gas concentration programming – a new approach to sample controlled thermal analysis. Thermochimica Acta 320, 297–301 (1998).
30. Barnes, P.A., Tiernan, M.J. and Parkes, G.M.B. Sample controlled thermal analysis – Temperature programmed reduction of bulk and supported copper oxide. Journal of Thermal Analysis and Calorimetry 56, 733–737 (1999).

31. Barnes, P.A., Parkes, G.M.B., Brown, D.R. and Charsley, E.L. Applications of new high resolution evolved gas analysis systems for the characterisation of catalysts using rate-controlled thermal analysis. Thermochimica Acta 269, 665–676 (1995).
32. Stacey, M.H. Constant rate thermal analysis for catalyst activation. Analytical Proceedings 22, 242–243 (1985).
33. Criado, J.M., Ortega, A. and Gotor, F. Correlation between the shape of controlled-rate thermal-analysis curves and the kinetics of solid-state reactions. Thermochimica Acta 157, 171–179 (1990).
34. Ortega, A., Akhouayri, S., Rouquerol, F. and Rouquerol, J. On the suitability of Controlled Transformation Rate Thermal-Analysis (CRTA) for kinetic-studies. 2. Comparison with conventional TG for the thermolysis of dolomite with different particle sizes. Thermochimica Acta 235, 197–204 (1994).
35. Vyazovkin, S. Two types of uncertainty in the values of activation energy. Journal of Thermal Analysis and Calorimetry 64, 829–835 (2001).
36. Reading, M., Dollimore, D., Rouquerol, J. and Rouquerol, F. The measurement of meaningful activation-energies – using thermoanalytical methods – a tentative proposal. Journal of Thermal Analysis 29, 775–785 (1984).
37. Tiernan, M.J., Barnes, P.A. and Parkes, G.M.B. New approach to the investigation of mechanisms and apparent activation energies for the reduction of metal oxides using constant reaction rate temperature-programmed reduction. Journal of Physical Chemistry B 103, 338–345 (1999).
38. Criado, J.M., Gotor, F.J., Ortega, A. and Real, C. The new method of Constant Rate Thermal-Analysis (CRTA) – application to discrimination of the kinetic-model of solid – state reactions and the synthesis of materials. Thermochimica Acta 199, 235–238 (1992).
39. Tiernan, M.J., Barnes, P.A. and Parkes, G.M.B. Reduction of iron oxide catalysts: The investigation of kinetic parameters using rate perturbation and linear heating thermoanalytical techniques. Journal of Physical Chemistry B 105, 220–228 (2001).
40. Dawson, E.A., Parkes, G.M.B., Barnes, P.A., Chinn, M.J. and Norman, P.R.A study of the activation of carbon using sample controlled thermal analysis. Journal of Thermal Analysis and Calorimetry 56, 267–273 (1999).
41. Real, C., Alcala, M.D. and Criado, J.M. Development of a new equipment for applying the Constant Rate Thermal-Analysis (CRTA) to Temperature Programmed Oxidation (TPO) of catalysis. Journal of Thermal Analysis 38, 797–802 (1992).
42. Dawson, E.A., Parkes, G.M.B., Barnes, P.A., Chinn, M.J. and Norman, P.R. Comparison of new thermal and reactant gas blending methods for the controlled oxidation of carbon. Thermochimica Acta 335, 141–146 (1999).
43. Fesenko, E.A., Barnes, P.A., Parkes, G.M.B., Dawson, E.A. and Tiernan, M.J. Catalysts characterisation and preparation using sample controlled thermal techniques – high resolution studies and the determination of the energetics of surface and bulk processes. Topics in catalysis 19, 291–309 (2002).

44. Farneth, W.E. and Gorte, R.J. Methods for characterizing zeolite acidity. Chemical Reviews 95, 615–635 (1995).
45. Salvador, F. and Merchan, M.D. Controlled-rate thermal desorption. Journal of Thermal Analysis and Calorimetry 51, 383–396 (1998).
46. Torralvo, M.J., Grillet, Y., Rouquerol, F. and Rouquerol, J. Application of CRTA to the study of microporosity by thermodesorption of preadsorbed water. Journal of Thermal Analysis 41, 1529–1534 (1994).
47. Chevrot, V., Llewellyn, P.L., Rouquerol, F., Godlewski, J. and Rouquerol, J. Low temperature constant rate thermodesorption as a tool to characterise porous solids. Thermochimica Acta 360, 77–83 (2000).
48. Fesenko, E.A., Barnes, P.A., Parkes, G.M.B., Brown, D.R. and Naderi, M. A new approach to the study of the reactivity of solid-acid catalysts: The application of constant rate thermal analysis to the desorption and surface reaction of isopropylamine from NaY and HY zeolites. Journal of Physical Chemistry B 105, 6178–6185 (2001).
49. Aramendia, M.A. et al. Comparison of different organic test reactions over acid-base catalysts. Applied Catalysis A-General 184, 115–125 (1999).

Chapter 8

SCTA IN THE FUTURE

M. READING

IPTME, Loughborough University, Loughborough, UK

8.1. Introduction

Many forms of SCTA and their applications have been described in the preceding chapters. In recent years, examples of the use of SCTA have been increasing and there would seem to be every reason to expect that this trend will continue. In this final chapter it is our purpose to discuss some of the new directions which SCTA might take as its advantages become clearer to a broader range of workers and as new technologies evolve that can usefully employ its principles. We will also outline some of the fundamental issues that still present challenges for the future.

8.2. New Sensors

The quantities that have featured most frequently in this book as controlling parameters in SCTA experiments include mass, gas flow, heat flow and length. There is scope for enhancing these methods by the use of new sensors. It is convenient to divide this discussion into two categories; new sensors that typically apply at the same scale as current methods, the macro scale, and those that are emerging in the new field of nanotechnology.

8.2.1. MACRO SCALE SENSORS

It will always be true that any new sensor that finds application in conventional thermal analysis can be adapted for SCTA. For mass measurements, the new vibrating cone device seems to offer an interesting new capability [1]. This technique is based on measuring the resonance frequency of a cone in which the sample is placed on a frit so that a gas stream can be passed through it. It is specially suited for SCTA experiments in which it is desirable to pass a gas through rather than above the sample, for example when using a reacting gas. It can be operated between vacuum and 60 bars, and between room temperature and 600°C.

For gas flow measurements, any new device for detecting gases either in the absence of other gases (partial vacuum) or in a vector gas could be of use for SCTA. The new electronic 'noses' might be present opportunities for SCTA [2] as might the vibrating tube gas density meter (made by the company Panasonic).

Heat flow measurements might be made using IR emission detectors or IR cameras directly measuring the temperature difference between a sample and it's surrounding as a real-time image.

The most interesting developments in the field of measuring dimensional changes are probably the emerging nano scale methods and these are discussed in the next section.

8.2.2. NANO SCALE SENSORS

The use of a quartz microbalance could extend the application of SCTA to thermogravimetry down to below 1 mg. It is based on the effect of the mass change on the resonance frequency of a large, flat, quartz crystal (on which the sample must be stuck). This apparatus has been used for sensitive gas adsorption studies [3] but has, to date, been more often used for electrochemical investigations [4]. For this latter application a commercial apparatus is currently available. There is a potential problem when using temperature as a variable because of the temperature dependence of the quartz resonating frequency. However this could be overcome by suitable calibration or the use of a differential set-up. Even more sensitive are miniature silicon cantilevers. In common with the quartz microbalances, their resonance frequency can be used to measure nanograms of material [5]. Micro machined cantilever technology can also be extended to differential thermal analysis. The degree of bending of a cantilever made from two materials with different coefficients of expansion can be a sensitive measure of temperature and, by extension, differential temperature [5].

There is the related field of micro-thermal analysis [6–8]. In this technique an active thermal probe is used in a scanning probe microscope. The thermal probe has an ultra miniature resistive heater at its tip so that it can be heated and, by measuring its resistance, temperature at the tip can be measured. A schematic diagram of one sort of probe of this type is given in Figure 8-1.

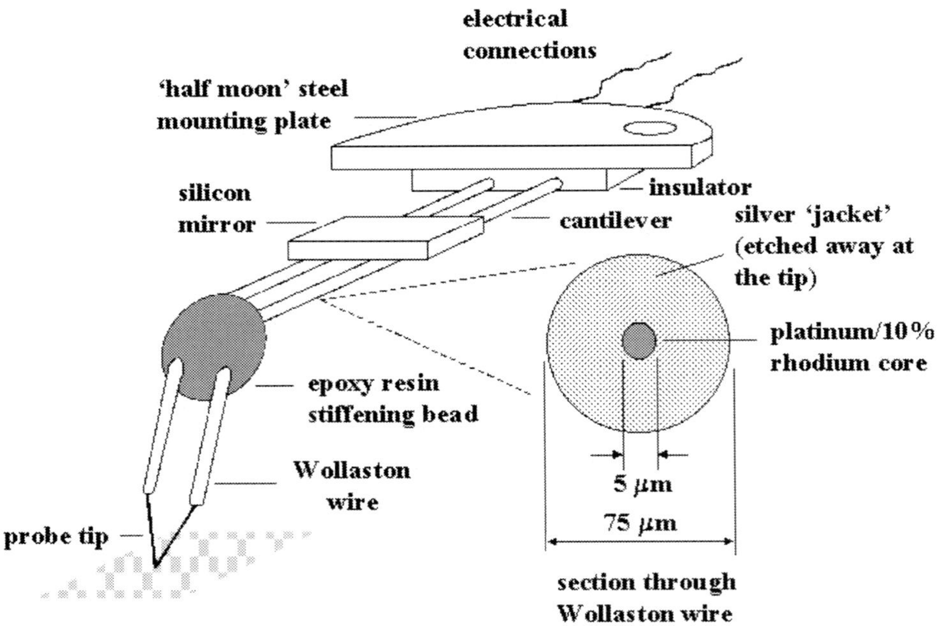

Figure 8-1. Schematic diagram of a thermal probe used for micro-thermal analysis.

In one mode of operation the temperature of the tip is held constant while the electrical power required to achieve this is measured. The probe can then be rastered over the surface of the sample to provide images of its relative thermal conductivity. It can also be placed on a selected point of interest on the surface and its temperature scanned to perform a local thermal analysis experiment. In this way local transition temperatures can be measured. One type of measurement is local thermomechanical analysis (L-TMA) where the indentation of the probe is measured. Another is local differential calorimetry (L-DSC) in which the sample probe is used in conjunction with a reference probe and the difference in electrical power supplied to the probes is measured. This measurement can also be used with a modulation to provide 3 calorimetric signals, the average signal and the amplitude and phase of the response to the modulation.

In this experiment the probe deflection provides the L-TMA information and the power signal, here shown as a derivative, is the L-DSC measurement (no modulation was used in this case). The melting point of the polymer can be clearly seen. The technique is useful for identifying and characterising phases in complex heterogeneous systems by determining local transition temperatures. A detailed discussion of this recently introduced technique is beyond the scope of this chapter and the reader is referred to [6–8] for further information. The question we address here is what benefits might there be in using SCTA approaches for these kinds of measurement? One possibility is that the heating rate is substantially increased just after the onset of a transition (detected, for example, as a downward movement in L-TMA). The reason why this might be beneficial is that heating too fast before the onset of a transition can compromise the accuracy of the temperature measurement. Once the transition has started there are benefits in using a fast heating rate because this causes less local damage. The zone that is heated by the experiment is greater when slower heating rates are used.

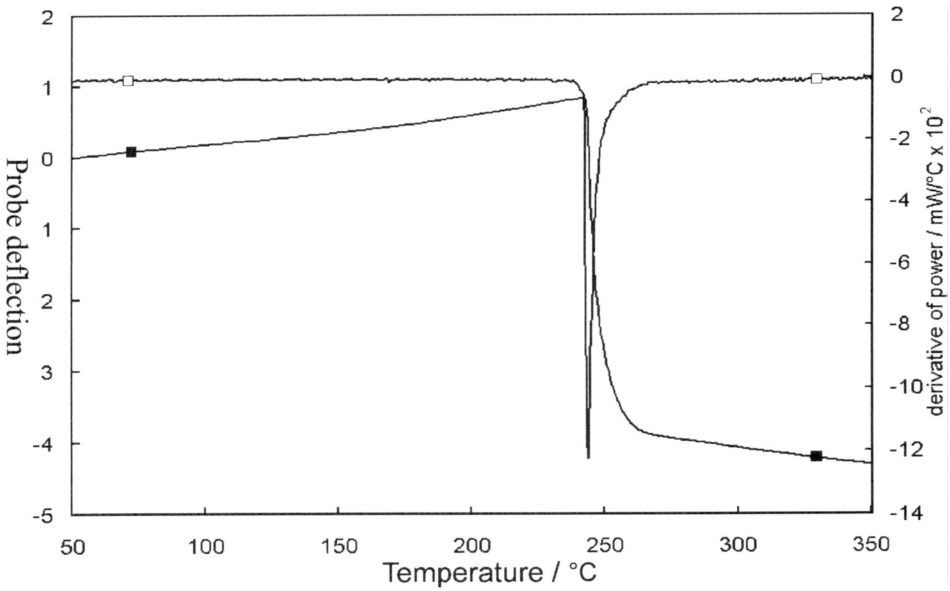

Figure 8-2. Example of a local thermal analysis experiment on a semi-crystalline polymer.

In another mode of use, these tips are used to cause local pyroloysis and the evolved gases are analysed by either mass spectroscopy or GC-MS. Here there is scope for applying SCTA ideas in the usual way; as the species are evolved the heating rate could be slowed to improve resolution between successive reactions. The total volume of material pyrolysis can be as small as a few tens of cubic microns. In the future this may decrease to cubic nanometers.

We have identified a very specific possible benefit when applying SCTA to micro-TA measurements, namely a reduction of the size of the zone affected by the measurement. If we look more generally and ask whether SCTA might be expected to provide benefits when applied to nano measurements, we might conclude that some of the principal benefits of SCTA, namely the reduction of pressure and temperature gradients, will not apply at this very small scale. However, one aspect of using ultra miniature systems is that the speed of response is often very fast and so experiments can be conducted very quickly. The whole experiment could last for a fraction of a second and this reintroduces potential problems of inhomogeneities within the sample bed. This then opens the door to the use of SCTA even on the nano scale. Only time will tell whether this will ultimately prove to be useful.

8.3. New Types of Measurements for SCTA

There remains scope for using many more types of already well established measurements for SCTA than have been used to date. These include methods for determining physical properties and analytical techniques that look at chemical composition.

Dimensional changes (a 'DC' mechanical property) have been used as a controlling parameter in SCTA but little attention has been paid to looking at dynamic mechanical (viscoelastic) properties or 'AC' mechanical properties. The first, and so far only, demonstration of this was by Reading [9]. He showed an apparent increase in resolution between two glass transitions measured in a blend of two polymers when compared with conventional linear heating. Care must be taken in interpreting this type of measurement because the dynamic modulus would, in many cases, not change with time. It is, in this sense, effectively a static quantity and thus fundamentally different from, for example, looking at rate of weight loss resulting from a chemical reaction because this is a process that progresses with time. Consequently, there is often a unique, barring experimental errors, plot of storage modulus (the real part of the dynamic modulus, see [9]) as a function of temperature that does not change with heating rate for transitions like glass transitions (which is where most applications for DMA lie). When this is the case, there must be a question mark over the relevance of SCTA for this application. The increased resolution seen in [9] probably came from decreasing the experimental errors in measuring the sample temperature in the region of interest. Problems of temperature gradients can be acute in DMA equipment because of the relatively large samples and massive clamps. The benefit that SCTA offers is essentially one of time saving. By heating rapidly when nothing is happening but slowly when a transition is in progress, good resolution (minimum temperature gradients) can be achieved in a shorter time. This is an advantage that can be particularly important in industrial laboratories. There might be further advantages in other cases where the dynamic modulus provides and an indirect measurement of the progress of some kinetic process like a chemical reaction, e.g. a curing reaction in a cross-linking polymer system or a change from one crystal form to another in an inorganic material.

The use of SCTA for calorimetry in the context of modulated temperature DSC (MTDSC) was briefly discussed in chapter 3. In the example outlined there, SCTA was used to create the modulation that was subsequently used as the basis of a deconvolution into reversing and non-reversing quantities in the way introduced by Reading [10–13]. We proposed that an interesting alternative would be to use a fixed frequency modulation, such as is usually used in MTDSC experiments, and use the non-reversing signal as the input to a SCTA algorithm. This is appropriate because this signal can serve as a measure of the progress of a chemical reaction in just the same way as a DTG measurement. The use of MTDSC can avoid the problem usually encountered when applying SCTA to calorimetry, namely that of contributions to the heat flow from not only the kinetic process, e.g. chemical reaction, but also the heat capacity. An alternative might be to use the reversing signal. When we look only at the reversing signal we have a technique equivalent to AC calorimetry [14] (in other words the total and non-reversing information are lost, in AC calorimetry only the amplitude and phase of the response to the temperature modulation are measured). Given that this is, in many cases, principally a measurement of the vibrational heat capacity, the application of SCTA to 'control' this quantity would seem to make little sense. The argument is similar to the one used above in respect of dynamic modulus when considering the DMA measurements (except the temperature gradients in a DSC, MTDSC or AC calorimetry experiment are typically small). If we take a glass transition (or series of glass transitions) then there is, to a good approximation, a unique plot of reversing heat capacity (and/or the phase lag) against temperature irrespective of the average heating or cooling rate. In this case SCTA would not offer any benefits. However, in cases where the reversing signal serves to indirectly measure the progress of, say, a chemical reaction then SCTA applied to the reversing signal (or AC calorimetry) might prove beneficial. Other possible examples could include a transition between different crystal forms where they have significantly different heat capacities or the loss of a volatile which would often give rise to a significant decrease in heat capacity.

The comments made above with regard to DMA, AC calorimetry and the reversing signal in MTDSC, apply to complex permittivity measured by dielectric thermal analysis (DETA) [15]. In this technique the sample is placed in an AC electrical field while its temperature is programmed. Perhaps more interesting are thermally stimulated current measurements (TSC) [16]. In this technique the relaxation of 'frozen' dipoles gives rise to an irreversible current that would certainly be susceptible to control using SCTA methods. The kinetics of these relaxations are of interest and the use of SCTA approaches could be beneficial especially when used with temperature (or rate) modulation (in fact modulation could be useful irrespective of the use of SCTA).

Another area of potential application of SCTA principles is that of acoustic measurements. Experiments in which sound waves are propagated through a sample are called thermoacousimetry [17]. The measured changes in propagation speed can be related to the sample's mechanical properties. This then brings us back to our discussion of dynamic modulus in DMA above. Thermosonimetry [18] is a technique in which the emission of sound is measured as the sample is heated. This method can detect pre-transition changes in samples undergoing solid-to-solid transitions. Because of this, when used in simultaneous combination with other measurements like DSC, it might be used to provide a warning that a transition is about to occur and so change the heating rate in anticipation of a transition occurring.

Although IR has already been used as a means of evolved gas analysis for SCTA [19], it has never been used as a direct measurement on a sample to control the progress of a chemical reaction. Any method of IR spectroscopy could be used, transmission, emission, photo acoustic, attenuated total internal reflectance etc. Similarly, Raman spectroscopy, nuclear magnetic resonance or any other form of spectroscopy could be used to measure and, through SCTA, influence the progress of a chemical transformation. X-ray diffraction could be used as the detector to control change in physical state. Other, as yet untried methods that might benefit from SCTA methods include emanation thermal analysis [20] and magnetic susceptibility measurements (Thermomagnetometry) [21].

In conclusion it can be said that all thermal methods with any form of detector *might* benefit from SCTA. However, in reality, there are many that would probably not benefit significantly. We can distinguish cases where a dynamic ('AC') measurement is used to plot changes in some quantity as a function of temperature and this plot is, to a good approximation (barring experimental errors) invariant with heating rate (although it may well vary with frequency). Examples include the dynamic modulus, reversing heat capacity and complex permittivity of glass transitions. SCTA approaches might be appropriate where these AC measurements indirectly measure the progress of an irreversible process such as the loss of a volatile material. However, this begs the question of whether these techniques are the most appropriate. Why not use a direct measure like thermogravimetry? Nevertheless, there may be cases where SCTA can be usefully applied to a purely AC measurement, perhaps when looking at solid-to-solid transitions. It is only by addressing this question experimentally that it can be resolved.

SCTA finds a more natural home when looking at transformations that are, at the time and temperature and under the experimental conditions they are made, irreversible. Examples include cure reactions in polymers or decomposition or desorption processes under high vacuum or in a vector gas. The measurements above that can follow these types of processes are all good candidates to benefit from SCTA approaches and we may well see these emerging in the near future.

8.4. Optimising Resolution and The Parameter Space Problem

Conventional linear rising temperature experiments require that the experimenter chose only one variable, the heating rate. Constant rate thermal analysis requires that the operator chose, usually, the rate of mass loss. The dynamic rate and maxres methods presented in chapter 3 present the user with several more parameters. These methods were specifically designed to maximise resolution (separation of successive transformations) and the problem is to find which parameters give the best balance between maximising resolution and minimising the duration of the experiment. Even with a relatively modest three parameters, the permutations quickly generate a large parameter space which leads to the generic problem of how to choose the best combination of variables. There is, in fact, no simple answer to this question. The use of an auto sampler can be very useful to carry out a series of experiments with different parameters. Experience is also, of course, a key to reducing the time taken to determine the optimum conditions. In general terms, however, this problem has no clear general solution.

8.5. Peak Shape Recognition

Most of the temperature programming methods described in chapter 3 can be summarised schematically as a plot of heating rate against rate of mass loss as shown in Figure 8-3 (where other measures of transformation rate could be substituted for rate of mass loss). The maxres method cannot so easily be described by this type of diagram but it shares an essential feature with these other methods. The response of the temperature controller is essentially governed by the value of some measure of the rate of transformation; this is exemplified by rate of mass loss in Figure 8-3. Peak shape recognition is intended to extend this idea by proposing that peak shape, rather than simply the rate of transformation, could be used as the basis of algorithms to improve resolution. The first proposal [22] is illustrated in Figure 8-4.

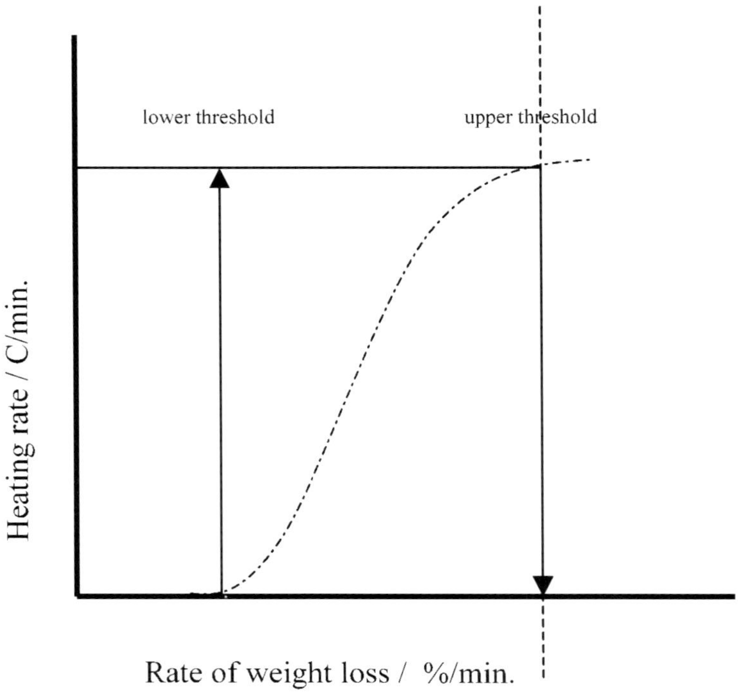

Figure 8-3. Composite schematic of various SCTA methods. The solid line represents the stepwise methods (that of Sorensen when the lower threshold is not zero and that of Honda when it is zero), the dashed vertical line represents the constant rate method while the dot-dash curve represents the dynamic rate method. Conventional linear heating rate methods can be illustrated as a horizontal line.

In the method demonstrated in Figure 8-4 a pre-selected heating rate is used until the first and second derivatives of the peak (from EGA or derivative of weight loss) are +ve and –ve at which point the heating rate is decreased. It is increased again when the first and second derivatives are –ve and +ve. Consider Figure 8-5, two possible results are presented for a two-step process. In the first of these two examples the peaks are of similar size, in the second example one of the peaks is much smaller. If the constant rate approach were adopted, a rate that would give a suitable result for the smaller peak would give an unnecessarily long experimental time for the bigger peak. A rate that might give a reasonable result for the larger process might give too short a duration for the smaller process. This peak shape algorithm resolves this problem because it would produce the same temperature profile for both peaks.

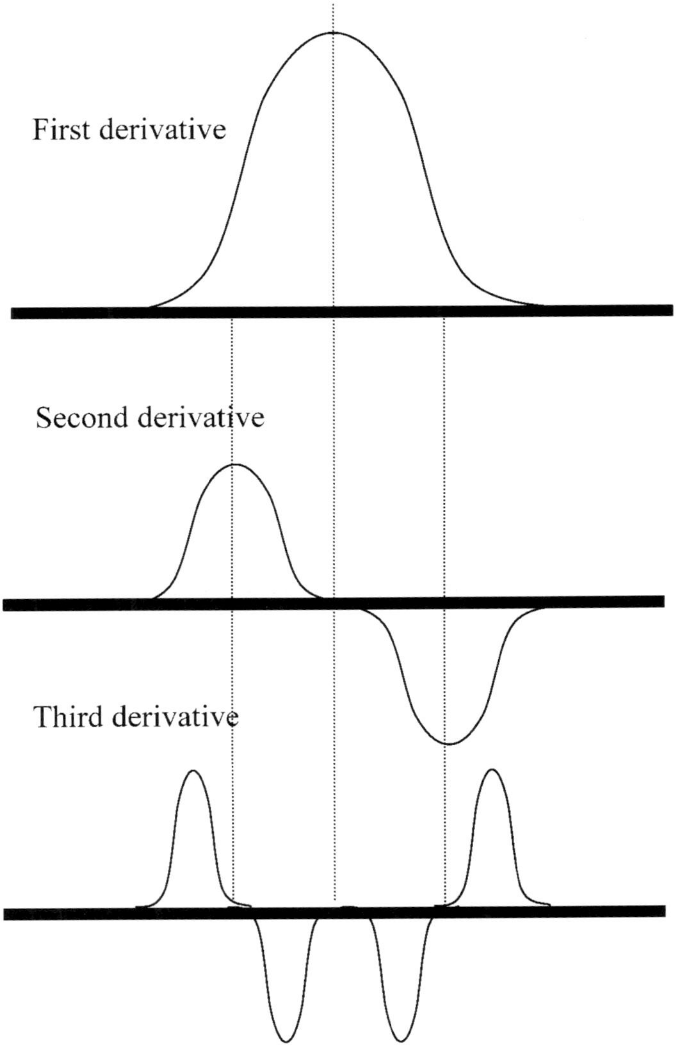

Figure 8-4. Illustration of the type of data used for a 'peak shape recognition' approach to SCTA.

It should be noted that this method is step-wise in character. Stepwise methods have the potential to provide much improved separation between successive reaction steps. However, there must never be more than one step within a single transformation otherwise it will appear to be at least two processes. This is the problem at the heart of stepwise methods applied to

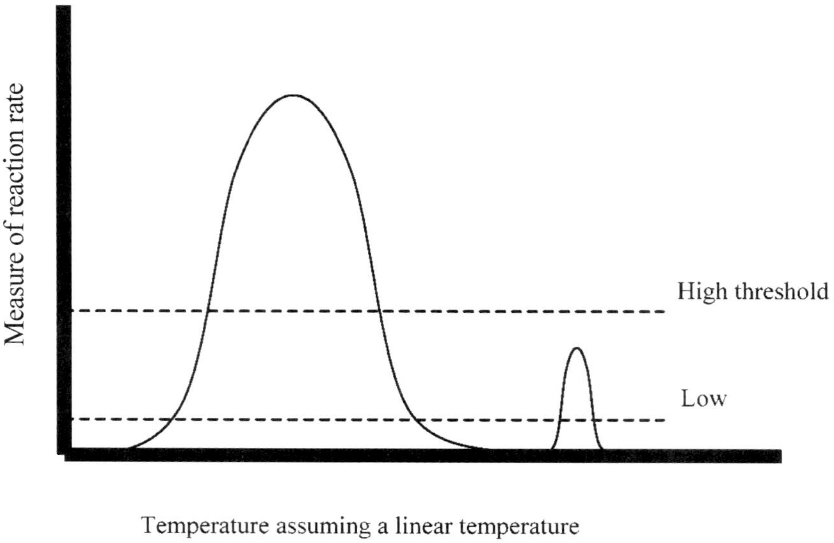

Figure 8-5. Illustration of the dilemma when choosing a reaction rate for a constant rate experiment with two processes of different magnitudes.

improving resolution. What choice of threshold values will avoid this difficulty and also avoid the problem of insufficient resolution so more than one transformation occurs during a single temperature step? The above 'peak shape recognition method' has the potential to avoid this problem because no threshold values for the reaction rate are selected. By the same token, it also potentially avoids the parameter space problem.

The interesting thing about the dynamic rate method illustrated in Figure 8-3 and discussed in detail in Chapter 3 (see 3.5.1) is that it provides a smooth and continuous (non-stepwise) relationship between transformation rate and heating rate. With this method neither rate of change of temperature nor transformation rate are maintained constant (see Fig. (8-3)-25) or follow some other predetermined function with time. This was the first non-stepwise method to adopt this kind of approach. Reading generalised this concept [8] and christened it "Constrained Rate Thermal Analysis" as a term for non-stepwise methods where neither the reaction rate nor the temperature follows a predetermined function of time. Peak shape recognition can also be adapted to provide for a non-stepwise algorithm. For example, the peak shape recognition approach was extended by this author and co-workers [23] using a concept based on anticipating the shape of the peak from the early data. If a Gaussian shape is assumed the total peak shape can be predicted from the early part of the curve and so the heating rate can be changed as a function of fractional conversion (using the total predicted area) or as a function of the standard deviation from the mid-point. In this way a continuous form of control could be used and a stepwise response avoided. Another approach could be to attempt to recognise the kinetic function. By way of illustration we can consider the reduced temperature approach of Reading [9,24].

Taking the conventional formalism for solid state kinetics viz:

$$d\alpha/dt = f(\alpha)\, Ae^{-Ea/RT} \tag{1}$$

Where α = fractional extent of conversion
$f(\alpha)$ = some function of extent of conversion
A = the pre-exponential constant
E_a = the activation energy
R = the gas constant
T = the absolute temperature

The reduced temperature is calculated as follows: Taking $T_{0.8}$ to be the temperature at α = 0.8 and $T_{0.3}$ to be the temperature at α = 0.3 (these being arbitrary choices), when reaction rate is held constant we may write:

$$(1/T_\alpha - 1/T_{0.9}) \,/\, (1/T_{0.3} - 1/T_{0.9}) = (\ln f(\alpha) - \ln f(0.8)) \,/\, \ln f(0.3) - \mathrm{Inf}(0.8) \tag{2}$$

$$(\mathrm{a} - T_\alpha) \,/\, \mathrm{a}T_\alpha \,/\, \mathrm{b} \;=\; (\ln f(\alpha) - \mathrm{q}) \,/\, \mathrm{d} \tag{3}$$

Where $\mathrm{a} = \mathrm{T}_{0.8}$
$\mathrm{b} = (T_{0.8} - T_{0.3}) \,/\, (T_{0.8}\, T_{0.3})$
$\mathrm{q} = \ln f(0.8)$
$\mathrm{d} = \ln f(0.3) - \ln f(0.8)$

Where a, b, q and d are constants. Figure 8-6 gives examples of 'master curves' for selected functions. Clearly this test cannot distinguish between different order expressions. If a reaction is found to follow an order expression, the exact order, n, can be established once a 'modulation' method has been used to determine the activation energy as detailed in previous chapters. While this approach is valid when the reaction rate is strictly constant, it does not apply when this condition is not met as in controlled rate and temperature methods (so called constrained rate methods) such as we are considering here. Reading went on to propose a variation on this method [9] where:

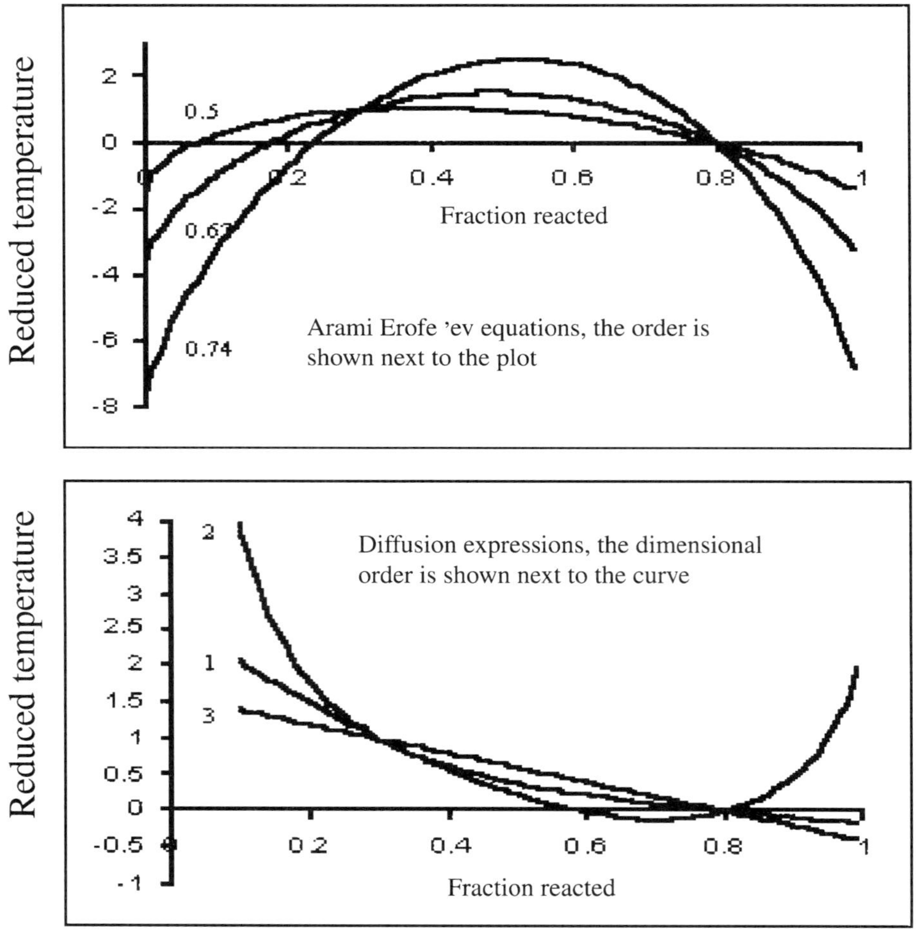

Figure 8-6. Reduced temperature master plots for various functions.

$$1/T_\alpha - 1/T_{0.8} \,/\, (1/T_{0.3} - 1/T_{0.8}) =$$
$$\ln[f(\alpha)/f(0.8))/(da/dt_\alpha/da/dt_{0.8})] \,/\ln[f(0.3)/f(0.8))/\,(da/dt_{0.3}/da/dt_{0.8})] \qquad (4)$$

where $da/dt_{0.8}$ = rate of reaction at $\alpha = 0.8$
$da/dt_{0.3}$ = rate of reaction at $\alpha = 0.3$

This cannot create true master curves but plots of the left hand side against the right hand side of Eq. (5) can be tried for different $f(\alpha)$ and the correct function should give a straight line with a gradient of 1 that passes through the origin [9]. This type of test is easily and rapidly carried out for a wide range of functions on modern computers and has been successfully applied to the dynamic rate method [9]. Let us consider a thermogravimetric experiment: if we restrict ourselves to order expressions we can reduce the number of unknowns in Eq. (5) at any point during a reaction to three. These are the mass at the start of a reaction step, the mass at the end of a reaction step and the order of the reaction n. In Eq. (5), α can be replaced by mass / (mass at the start of reaction step – mass at the end of reaction step) and $d\alpha/dt$ can be replaced by rate of mass loss /(mass at the start of reaction step – mass at the end of reaction step), $f(\alpha)$ can be replaced by $(1 - \alpha)^n$ where n is the reaction order. It must be remembered that the choice of 0.3 and 0.8 values for extent of reaction is arbitrary; any two points can be selected. A judicious choice in real time at any point during an experiment might be the most recent values for mass and rate of mass loss and 50% of this value where 0% is the starting mass or the mass at the end of the proceeding reaction step as delineated by the point at which the heating rate was increased (this could also be allowed as a variable in the fitting procedure within certain limits). The question is then which values for the weight losses and n give a best fit to the full data set. There are a number of multi-parameter fitting methods that could be used to estimate the three (possibly four) unknowns, or a neural net might be trained using 'ideal' theoretical data. Once successive values appear consistent then this provides a prediction for the shape of the TGA curve. This can then be used as the basis for an algorithm that progressively decreases the heating rate in anticipation of the maximum rate of reaction (rather than simply responding to the event when it occurs) and that increase the heating rate once a high percentage of the expected weight loss has been achieved. Alternatively, once the peak shape has been predicted with sufficient confidence from the initial part of the reactions step, this prediction could be used to select the parameters for either the dynamic rate or maxres algorithms (and so this become a way of resolving the parameter space problem). These same general arguments can be applied to more complex approaches that allow a wider variety of candidate kinetic functions.

These examples of possible peak shape recognition approaches are only intended to be illustrative. The basic thrust of the arguments presented above is that control algorithms no longer have to be based on selecting values for heating rates and rates of transformation (as expressed by some instrumental measurement). The use of computer base control systems greatly increases the flexibility and sophistication of the approaches that can be adopted. Regrettably, none of these peak-fitting approaches have yet been tried in practice. Hopefully this might change in the future.

8.6. Which Algorithm for Best Resolution?

It is probably true that most people in most applications are more interested in achieving good resolution so they can distinguish the start and end of different processes and determine from, for example, weight measurements, the amounts of different materials. Consider, for simplicity, the case of a thermogravimetric experiment that measures two independent processes that give rise to two distinguishable but overlapping steps at 10 C/min, each reaction having a given set of kinetic parameters. If it is allowed that the experiment can take 1 hour, is there a unique answer to the question "what temperature program will achieve the best separation possible of these 2 events?" This then begs the question, how do we define the 'best' resolution. One answer would be that the amount of material associated with the first reaction should have reached the minimum possible value before the second process starts and that we are able to distinguish the point at which this occurs. It is important that we can see when this point is reached so that we can quantify the amount of each material with the maximum accuracy. If there is a unique answer to the temperature program question, the next questions are "is there an algorithm that can take the weight loss as an input and control the furnace in real time that will find this temperature program?" and then "is it robust so that noise etc. won't give rise to artefacts that will give misleading outcomes (artefacts can occur when the demands made on the control system are too great)?" If there is no unique solution and/or no algorithm that will automatically derive it and/or the algorithm is not robust then we must ask "which control algorithm provides the best approximation to the ideal temperature program and also provides the best compromise between resolution and robustness" (we will avoid the problem of defining robustness and assume the reader will allow that there must be a system-dependent pragmatic resolution of this question). Is it the Honda method, linear temperature ramp, constant rate, dynamic rate (with some set of parameters), maxres (with some set of parameters), some variation of the peak shape recognition method or something else that will provide the best results? If it is the dynamic rate or maxres methods, how do we find the right settings (the parameter space problem referred to above)? Are there, in fact, no real answers to these questions because it is very dependent on the nature of the sample?

It would appear that there are currently no clear theoretical answers to this conundrum and before we have a theoretical solution there can be no practical solution. This would seem to be one for the biggest remaining problems in the field of sampled controlled thermal analysis.

8.7. SCTA when Dealing with Multiple Parameters

In conventional thermal methods the use of multiple detectors that each provides measurements of a different property of the sample (or a single detector that provides multiple measurements such as mass spectroscopy) poses no special problem. The desired temperature program is selected and the outputs from the various measurement devices are recorded as a function of time while the sample temperature changes as a predetermined function of time. This is most commonly employed when a reaction gives rise to multiple chemical species and mass spectroscopy, or some other type of EGA, is used in an attempt to measure the evolution of each of the various products.

When using SCTA, the use of multiple measurements does pose a question; "which of the measurements (corresponding to which chemical species or combination of chemical species) should be used as the control parameter?" SCTA using EGA where the rate of evolution of a single chemical species is controlled has been realised by Stacey [25] using a dew-point hygrometer and a catharometer, by Reading at al. using twin IR gas analysers [19] and by Rouquerol et al. using a mass spectrometer [26]. Typically when using a mass spectrometer the total ion current is used as the controlled parameter, however, a single ion can be used and, in simple cases, this can correspond to a single species.

To exemplify the kind of problems that can arise in practice we can consider the IR method of Reading [19,27] applied to cobalt hydroxycarbonate [28], a schematic of the apparatus is given in Figure 8-7 and the results are shown in Figure 8-8. The control system 'wanted' to increase the temperature very rapidly at the start of the experiment in order to cause carbon dioxide to evolve. The presence of a great deal of water, both as H_2O and as the hydroxide, meant that a great deal of moisture was evolved very rapidly and this initially defeated the control system. Consequently the experiment started in an uncontrolled fashion and took some time before proper control was established (and so the linear plot for CO_2 does not start at zero). The simplest solution would probably have been to start with a linear rising temperature program that 'handed over' control to the constant rate regime when the set partial pressure was reached. An alternative would have been to start with constant rate control of water then switch to carbon dioxide as soon as this gas reached a certain level. An extreme example of this type of practical problem is an innovative example of switching dilatometry and thermogravimetry [29] in an apparatus constructed by Thomas et al..

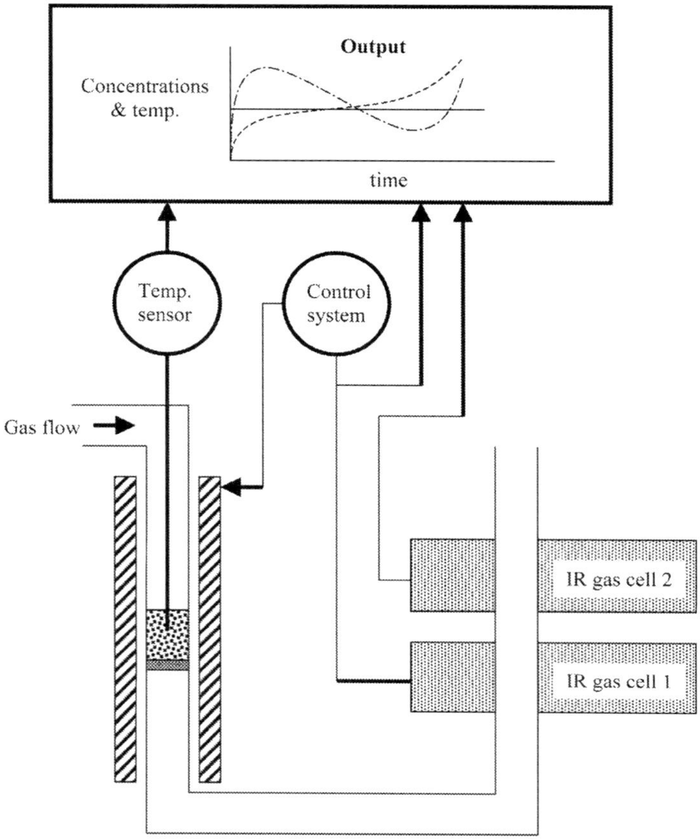

Figure 8-7. Schematic of the IR based constant rate apparatus constructed by Reading.

Now that a vast range of possibilities has been opened up by computer control we are forced to ask much more than in the past, which options result in the best experiments. It is not clear that any general guidelines can be offered for cases where more than one measured parameter could be of importance. When dilatometry and thermogravimetry were used [29] an approach was adopted that meant either the weight loss or dimension change could control the temperature depending on which signal was changing most. The objective was to control both to avoid damage to the sample. In the example where SCTA was applied to DMA [9], the rate of change of the amplitude of the force modulation was used as the controlled parameter whereas either the loss modulus or phase angle would have been a better choice if secondary transitions had been of interest because these signals change more during these types of processes. When multiple chemical species are being detected, such as in the case of IR [27,28] and MS [26], simple cases where two species are evolved can be dealt with in the manner discussed above but when multiple products are involved, looking at total ion current or total IR absorbance means individual species are uncontrolled and the implications of this on a kinetic or mechanistic interpretation could well be problematic. Controlling the rate of evolution of a single species might be impossible, depending on the complexity of the reaction and the detector being used, and where possible can pose the kind of practical problems illustrated in Figure 8-8 as well as problems of interpretation because the pressure of other species is not maintained constant.

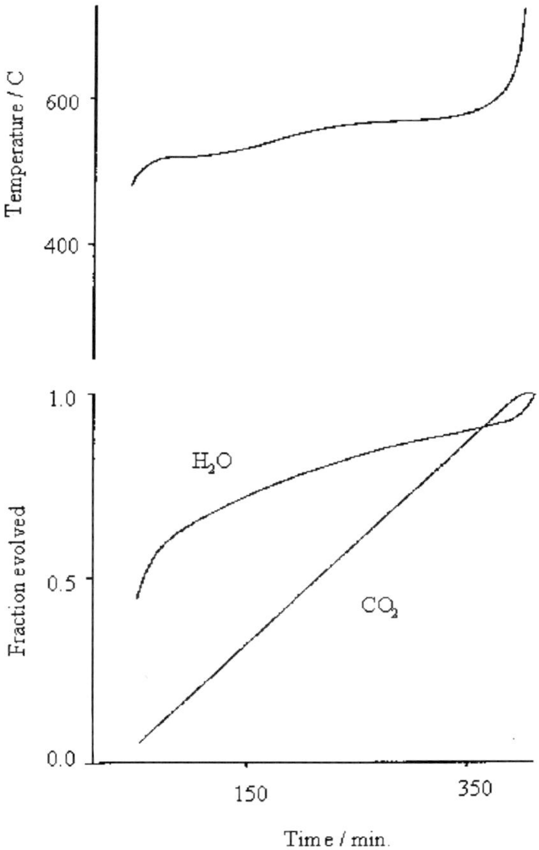

Figure 8-8. Example of results obtained with the apparatus illustrated in Figure 8-7 on cobalt hydroxycarbonate. The rate of evolution of carbon dioxide was controlled as constant while the amount of water evolved was simultaneously monitored.

SCTA, therefore, poses special problems when looking at multiple parameters simultaneously. In conventional thermal analysis the problem is greatly simplified. The temperature follows the chosen program and the many parameters are measured. The experimenter is not forced to ask him or herself any difficult questions. In SCTA, the question of which parameter or combination of parameters should be used and which control algorithm must be asked. Currently, in common with many problems in characterisation science, it must be left to the skill and ingenuity of the experimenter to design the best experiments and make the most meaningful interpretation possible within the constraints of the equipment available to him. However, we can expect that multiparameter experiments will become more common in the future as equipment becomes more sophisticated. A lack of generic solution to the problem of dealing with these types of experiment will be an impediment to the wider acceptance of SCTA methods. A challenge for the future will be to invent appropriate procedures for the most common types of analysis.

8.8. Overview

SCTA has seen an explosion of activity in the past decade in large part due to the introduction of commercially available versions of this approach. In the past it has been the preparation of porous solids and the study of kinetics that have been the drivers for the development of these techniques, more recently it is probably improved resolution of overlapping events. Today the experimenter is faced with a much broader range of techniques than in the past together with the questions that arise from this variety. To which techniques is SCTA most profitably applied? If EGA is chosen, which approach should be used both in terms of the method of analysing the gases and the algorithm that should be used to translate the measurements into some temperature program? More generally, how should the question of control regimes for multiparameter measurements be tackled? The early promise CRTA showed as a means of studying kinetics remains, in this author's view, tantalisingly encouraging and yet fundamental questions regarding solid-state decomposition kinetics remain unanswered. At the heart of SCTA is the control algorithm. What should it be in order to answer a given question even when only a single parameter is being is being used as the input to the control system?

The modern literature, of which this book is part, contains a lot of information that should, in many cases, be able to lead the informed researcher to valid conclusions on how best to use SCTA in their research. There are many exciting possibilities for the future and every reason to believe the benefits described in this book can be extended to other instruments and fields of application. Notwithstanding this, some fundamental questions remain open and so those engaged in developing the techniques themselves still have challenges in front of them.

References

1. H. Patashnik, G. Rupprecht and J.C.F. Wang, *Prepr. Am. Chem. Soc., Div. Pet. Chem.*, 25 (1980) 188. This device is available under the name of Tapered Element Oscillating Microbalance (TEOM).
2. B.J. Doleman, E.J. Severin and N.S. Lewis, Proc. National Acad. Sciences, 95 (1998) 5442–5447.
3. J. Krim and E.T. Watts, in *Third International Conference on Fundamentals of Adsorption*, A.B. Mersmann and S.E. Scholl (Eds.), Engineering Foundation, New York (1991) 445.
4. M.R. Deakin and O.R. Melroy, *J. Electrochem. Soc.*, Vol. 136, No. 2 (1989) 349. This device is commercially available known as the Electrochemical Quartz Crystal Microbalance (EQCM).
5. R. Berger, Ch. Gerber, H.P. Lang and J.K. Gimzewski, *Microelectronic Engineering*, 35 (1997) 373–379.
6. A. Hammiche, M. Reading, H.M. Pollock, Mo Song and D.J. Hourston, *Review of Scientific Instrumentation*, 6712 (1996) 4268–4273.
7. M. Reading, D.M. Price, D. Grandy, R.M. Smith, L. Bozec, M. Conroy, A. Hammiche and H.M. Pollock, *Macromolecular Symposia*, 167 (2000) 45–62.
8. H.M. Pollock and A. Hammiche, *J. Phys D: Appl. Phys.* 34 (2001) R23–R53.
9. M Reading, *Thermal Analysis – Techniques and Applications*, (Ed.) E.L. Charsley and S.B. Warrington, Chapter 7 (1992) 126.
10. M. Reading, D. Elliott and V.L. Hill, *Proc. NATAS* (1992) 145.
11. M. Reading, *Trends in Polymer Science*, 1 (1993) 8.
12. A.A. Lacey, C. Nikolopoulos and M. Reading, 50 (1997) 279.
13. P. Haines, C. Keattch and M. Reading, Differential Scanning Calorimetry, *Handbook of Thermal Analysis*, (Ed.) M. Brown, Elsevier (1998).
14. Y.A. Kraftmakher, *Zh. Prikl. Mekh. & Tekh. Fiz.,* 5 (1962) 176.
15. V.V. Daniel, *Dielectric Relaxation*, Academic Press, New York, 1967.
16. A. Lamure, N. Hittini, C. Lacabanne, M.F. Herdmand and D. Herbage, *IEEE Trans. El.*, 21 (1986) 443.
17. K. Lonvik, *J. Therm. Anal.*, 25 (1982) 109.
18. T. Mraz, K. Rajeshwar and J. Dubow, *Thermochim. Acta*, 38 (1980) 389.
19. M. Reading and J. Rouquerol, *Thermochim. Acta*, 85 (1985) 299.
20. V. Balek and J. Tolgyessy, *Emanation Thermal Analysis and other Radiometric Emanation Methods*, in (Ed.) G. Svehla, Wilson and Wilson's *Comprehensive Analytical Chemistry*, Part XIIC, Elsevier, Amsterdam (1984).
21. P.K. Gallagher, *J. Thermal Anal.*, 49 (1997) 33.

22. M. Reading, Constant Rate Thermal Analysis and Related Techniques, *Handbook of Thermal Analysis*, (Ed.) M. Brown, Elsevier (1998).
23. G.M.P. Parkes, P.A. Barnes, E.L. Charsley, M. Reading and I. Abrahams, *Thermochim Acta*, 354 (2000) 39.
24. M. Reading, *Thermochim. Acta*, 135 (1988) 37.
25. M.H. Stacey, *Proc. 2nd European Sym. On Thermal Anal.*, (Ed.) D. Dollimore, Hayden, Aberdeen (1981) 480.
26. G. Thevand and F. Rouquerol, *J. Thermal Anal.*, 13 (1971) 413.
27. M. Reading and D. Dollimore, *Thermochim. Acta*, 240 (1994) 117–127.
28. M. Reading, PhD Salford UK 1983.
29. M.E. Thomas, S.P. Terblanche, J.W. Stander, K. Gilbert, L.P. Nortman and N.A. Stone, *Proceedings of the International Ceramics Conference 94, International Ceramics Monographs*, (Eds.) C.C. Sorell and A.J. Ruys, 1 (1994) 281.

Subject Index

Hot Topics in Thermal Analysis and Calorimetry

Series Editor:
Judit Simon, *Budapest University of Technology and Economics, Hungary*

1. Michael E. Brown: *Introduction to Thermal Analysis, Second Edition*. 2001
   ISBN 1-4020-0211-4; Pb 1-4020-0472-9
2. W. Zielenkiewicz and E. Margas: *Theory of Calorimetry*. 2002
   ISBN 1-4020-0797-3

KLUWER ACADEMIC PUBLISHERS – DORDRECHT / BOSTON / LONDON